Optics: A Formula Handbook

N.B. Singh

DEDICATION

To Nature,

I dedicate this book to you, the source of all life. You are my inspiration, my teacher, and my friend.

Thank you for teaching me about the beauty of the world around me. Thank you for showing me the power of the natural world. Thank you for giving me a sense of peace and tranquillity.

I promise to do my part to protect you and your many wonders. I will teach my children about the importance of conservation and sustainability. I will work to make the world a better place for all living things.

Thank you for everything, Nature.

With love,

N.B Singh

Contents

III   Light Sources   41

6 Light Emission   43

7 Lasers and Coherent Sources   49

Preface

Introduction

Welcome to "Optics: A Formula Handbook." This handbook serves as a comprehensive guide to the fundamental principles, essential formulas, and practical applications in the field of optics. Designed to be a valuable resource for students, researchers, and professionals, this book covers a wide range of topics, from basic optical concepts to advanced applications.

Objective

The primary objective of this handbook is to provide a concise yet thorough compilation of formulas, equations, and relevant information in optics. It aims to assist readers in understanding and applying optical principles to solve problems, design optical systems, and explore cutting-edge technologies.

Structure of the Handbook

The handbook is organized into several parts, each focusing on a specific aspect of optics. The content is structured to facilitate easy navigation and quick reference. The parts include:

1. **Fundamentals of Optics:** Basic concepts, laws of optics, and introductory principles.

2. **Wave Optics:** In-depth coverage of wave properties of light, interference, and diffraction.

3. **Optical Systems:** Understanding and designing optical components and systems, including lenses and mirrors.

4. **Light Sources:** Comprehensive information on various light sources, including incandescent, gas discharge, semiconductor, and lasers.

5. **Optics in Practice:** Practical aspects such as optical materials, coatings, and imaging systems.

6. **Advanced Topics:** Exploration of quantum optics, nonlinear effects, and emerging technologies.

7. **Appendices:** Additional resources, glossary, and mathematical formulas.

Who Should Read This Handbook?

This handbook is designed for a diverse audience, including:

- Students studying optics at the undergraduate or graduate level.

- Researchers and scientists in the field of optics and photonics.

- Engineers and professionals involved in optical system design and development.

- Anyone seeking a quick reference for optical formulas and principles.

How to Use This Handbook

Feel free to navigate through the handbook based on your specific needs. Whether you are looking for a quick formula, exploring advanced topics, or learning optical principles from scratch, this handbook provides a structured and accessible resource.

Disclaimer

This handbook is intended for educational and reference purposes. The author and publisher are not responsible for any errors or omissions or any consequences arising from the use of the information presented herein.

Part I

Introduction to Optics

Chapter 1

Basics of Optics

1.1 Overview of Optics

Optics is the branch of physics that deals with the study of light and its interactions with matter. Understanding optics is crucial in various scientific and technological fields, including astronomy, microscopy, telecommunications, and more. This section provides a brief overview of the fundamental concepts in optics.

1.1.1 Nature of Light

Light is often described as electromagnetic waves, propagating through space. The wave nature of light is characterized by its wavelength (λ) and frequency (f), related by the speed of light (c) through the equation $c = \lambda f$. This fundamental relationship forms the basis for understanding various optical phenomena.

1.1.2 Historical Perspective

The study of optics has a rich history that dates back to ancient civilizations. Early scholars, such as Euclid and Ptolemy, contributed to the understanding of basic optical principles. The development of lenses and the understanding of refraction by Ibn al-Haytham (Alhazen) in the 11th century laid the groundwork for further advancements. The works of Kepler and Newton in the 17th century further shaped the field, leading to the modern science of optics.

1.1.3   Sample Working Example: Snell's Law

One fundamental concept in optics is Snell's Law, which describes the relationship between the angles of incidence and refraction when light passes through different media. Snell's Law is given by:

$$n_1 \sin(\theta_1) = n_2 \sin(\theta_2) \tag{1.1}$$

where n_1 and n_2 are the refractive indices of the two media, and θ_1 and θ_2 are the angles of incidence and refraction, respectively.

1.1.4   Numerical Example: Calculating Refraction Angle

Consider light traveling from air ($n_1 = 1.0$) to glass ($n_2 = 1.5$) with an incidence angle of $\theta_1 = 30°$. Using Snell's Law, we can calculate the refracted angle θ_2:

$$1.0 \sin(30°) = 1.5 \sin(\theta_2) \tag{1.2}$$

$$\theta_2 = \sin^{-1}\left(\frac{1.0}{1.5} \sin(30°)\right) \tag{1.3}$$

$$\theta_2 \approx 19.47° \tag{1.4}$$

Thus, the refracted angle is approximately $19.47°$.

1.2   Nature of Light

The nature of light is a fundamental aspect of optics, and it is classically described as an electromagnetic wave. This wave nature is characterized by certain properties that play a crucial role in understanding optical phenomena.

1.2.1 Wave Characteristics

Light exhibits wave characteristics, including wavelength (λ) and frequency (f). The speed of light (c) is related to these properties by the equation $c = \lambda f$. This relationship is fundamental to understanding the behavior of light waves.

1.2.2 Wave-Particle Duality

While the wave model effectively describes many optical phenomena, light also demonstrates particle-like behavior. This duality is a cornerstone of quantum mechanics. The quantization of light into discrete packets called photons is described by the equation $E = hf$, where E is the energy of the photon, h is Planck's constant, and f is the frequency.

1.2.3 Polarization of Light

Light waves can be polarized, meaning the oscillations occur in a specific direction. This phenomenon is crucial in applications such as glare reduction and 3D movie technology. The polarization state of light is represented using the Jones matrix formalism.

1.2.4 Sample Working Example: Double-Slit Interference

A classic example showcasing the wave nature of light is the double-slit interference experiment. When light passes through two closely spaced slits, an interference pattern emerges on a screen. The fringe separation (Δy) is given by the equation:

$$\Delta y = \frac{\lambda L}{d} \tag{1.5}$$

where λ is the wavelength, L is the distance to the screen, and d is the slit separation.

1.2.5 Numerical Example: Interference Pattern

Consider a double-slit experiment with a red laser ($\lambda = 650$ nm), a screen 2 m away, and slits separated by 0.1 mm. The fringe separation is calculated as:

$$\Delta y = \frac{(650 \times 10^{-9} \text{ m})(2 \text{ m})}{0.1 \times 10^{-3} \text{ m}} \tag{1.6}$$

$$\Delta y \approx 13 \text{ mm} \tag{1.7}$$

Thus, the interference pattern exhibits fringes approximately 13 mm apart.

1.3 Historical Perspective

The historical development of optics spans centuries, with contributions from various cultures and thinkers shaping our understanding of light and its properties.

1.3.1 Ancient Contributions

Optical concepts have roots in ancient civilizations. The Greeks, notably Euclid, explored the principles of reflection and proposed the idea that light travels in straight lines. The Romans later expanded on these ideas. However, it was during the Islamic Golden Age that significant advancements occurred. Ibn al-Haytham (Alhazen) conducted groundbreaking experiments on optics, including studies on refraction and the camera obscura.

1.3.2 Medieval Europe

In medieval Europe, the understanding of optics continued to evolve. The development of eyeglasses in the 13th century marked a practical application of optical principles. Visionaries like Roger Bacon and John Pecham contributed to the theoretical aspects of optics, and the magnifying properties of lenses were explored.

1.3.3 Kepler's Contribution

Johannes Kepler, in the early 17th century, provided crucial insights into the nature of light. He proposed that lenses could focus light, laying the groundwork for the development of telescopes. Kepler's work on the laws of planetary motion and optics influenced subsequent generations of scientists.

1.3.4 Newton and the Particle Theory

Isaac Newton, in the late 17th century, challenged the prevailing wave theory of light with his particle theory. He argued that light consists of particles, or corpuscles, and demonstrated that a prism could decompose white light into its constituent colors. Newton's corpuscular theory dominated for some time, but the wave theory was later revived.

1.3.5 Wave Theory Resurgence

The 19th century witnessed the resurgence of the wave theory of light. Thomas Young's double-slit experiment demonstrated interference, supporting the wave nature of light. Augustin-Jean Fresnel's work on diffraction and polarization further strengthened the wave theory, ultimately leading to the acceptance of the wave-particle duality.

1.3.6 Maxwell's Electromagnetic Theory

James Clerk Maxwell's formulation of the electromagnetic theory of light in the mid-19th century provided a unified understanding of electricity, magnetism, and light. Maxwell's equations predicted the existence of electromagnetic waves and paved the way for the development of technologies like radio waves and lasers in the 20th century.

1.3.7   Sample Working Example: Newton's Prism Experiment

Newton's experiment with a prism showed that white light could be separated into its component colors. The angular deviation ($\Delta\theta$) of each color is given by the equation:

$$\Delta\theta = \arcsin\left(\frac{n(\lambda) - 1}{\sqrt{2n(\lambda)}}\right) \tag{1.8}$$

where $n(\lambda)$ is the refractive index of the prism material for a given wavelength λ.

1.3.8   Numerical Example: Dispersion in Prism

Consider a glass prism with a refractive index of 1.5 for violet light ($\lambda = 400$ nm). Using the dispersion formula, the angular deviation is calculated as:

$$\Delta\theta = \arcsin\left(\frac{1.5 - 1}{\sqrt{2 \times 1.5}}\right) \tag{1.9}$$

$$\Delta\theta \approx 42.55° \tag{1.10}$$

Thus, violet light is deviated by approximately 42.55°.

Chapter 2

Geometrical Optics

2.1 Reflection and Refraction

The study of reflection and refraction is essential in understanding how light interacts with surfaces and interfaces. Geometrical optics provides a framework to analyze these phenomena using simple geometric principles.

2.1.1 Laws of Reflection

The law of reflection states that the angle of incidence (θ_{in}) is equal to the angle of reflection (θ_{re}) when a light ray reflects off a surface. Mathematically, this is expressed as:

$$\theta_{in} = \theta_{re} \tag{2.1}$$

This fundamental principle is crucial in understanding how light reflects off mirrors and other surfaces.

2.1.2 Laws of Refraction (Snell's Law)

When light passes from one medium to another, it changes direction. Snell's Law describes this phenomenon, stating that the ratio of the sines of the angles of incidence and refraction is equal to the ratio of the speeds of light in the two media. Mathematically, Snell's Law is given by:

$$n_1 \sin(\theta_{\text{in}}) = n_2 \sin(\theta_{\text{re}}) \tag{2.2}$$

where n_1 and n_2 are the refractive indices of the first and second media, respectively.

2.1.3   Sample Working Example: Reflection off a Plane Mirror

Consider a light ray incident on a plane mirror at an angle of 30°. According to the law of reflection, the reflected ray will also make an angle of 30° with the normal. This simple example illustrates the application of the law of reflection.

2.1.4   Numerical Example: Refraction in a Glass Prism

Imagine a light ray passing from air ($n_1 = 1.0$) into a glass prism ($n_2 = 1.5$) at an angle of incidence of 45°. Applying Snell's Law, we can calculate the angle of refraction:

$$1.0 \sin(45°) = 1.5 \sin(\theta_{\text{re}}) \tag{2.3}$$

$$\theta_{\text{re}} = \sin^{-1}\left(\frac{1.0}{1.5} \sin(45°)\right) \tag{2.4}$$

$$\theta_{\text{re}} \approx 30° \tag{2.5}$$

Thus, the refracted angle in the glass prism is approximately 30°.

2.1.5   Total Internal Reflection

When light travels from a medium with a higher refractive index to one with a lower refractive index, there is a critical angle beyond which total internal reflection occurs. This phenomenon is exploited in fiber optics and other optical devices.

2.1.6 Matrix Representation of Reflection and Refraction

Reflection and refraction can be conveniently represented using matrices. The reflection matrix for a plane mirror is:

$$R = \begin{bmatrix} 1 & 0 \\ 0 & -1 \end{bmatrix} \tag{2.6}$$

The refraction matrix for a medium with refractive index n is:

$$T = \begin{bmatrix} 1 & 0 \\ 0 & \frac{1}{n} \end{bmatrix} \tag{2.7}$$

These matrices are useful in analyzing the transformation of light rays at interfaces.

2.2 Lenses and Mirrors

Lenses and mirrors are fundamental optical components that play a crucial role in image formation and manipulation. Understanding their properties and behaviors is essential in the field of geometrical optics.

2.2.1 Lenses

Lenses are transparent optical elements with curved surfaces that can converge or diverge light. There are two main types of lenses: converging (convex) lenses and diverging (concave) lenses.

Lens Equation

The lens equation relates the focal length (f), object distance (d_{obj}), and image distance (d_{img}) for a lens. It is given by:

$$\frac{1}{f} = \frac{1}{d_{\text{obj}}} + \frac{1}{d_{\text{img}}} \tag{2.8}$$

Magnification

The magnification (m) produced by a lens is given by the ratio of the image height (h_{img}) to the object height (h_{obj}):

$$m = \frac{h_{\text{img}}}{h_{\text{obj}}} \tag{2.9}$$

Sample Working Example: Converging Lens

Consider a converging lens with a focal length of 10 cm. If an object is placed 20 cm from the lens, we can use the lens equation to calculate the image distance:

$$\frac{1}{f} = \frac{1}{d_{\text{obj}}} + \frac{1}{d_{\text{img}}} \tag{2.10}$$

Substituting values, we get:

$$\frac{1}{10 \text{ cm}} = \frac{1}{20 \text{ cm}} + \frac{1}{d_{\text{img}}} \tag{2.11}$$

Solving for d_{img}, we find $d_{\text{img}} = 20$ cm. Thus, the image is formed 20 cm from the lens.

Numerical Example: Magnification of a Diverging Lens

For a diverging lens with a magnification of -0.5, if the object height is 4 cm, we can find the image height using the magnification formula:

$$m = \frac{h_{\text{img}}}{h_{\text{obj}}} \tag{2.12}$$

Substituting values, we get:

$$-0.5 = \frac{h_{\text{img}}}{4 \text{ cm}} \tag{2.13}$$

Solving for h_{img}, we find $h_{\text{img}} = -2$ cm. The negative sign indicates an inverted image.

2.2.2   Mirrors

Mirrors, like lenses, can also form images. There are two main types of mirrors: concave mirrors and convex mirrors.

Mirror Equation

The mirror equation relates the focal length (f), object distance (d_{obj}), and image distance (d_{img}) for a mirror. It is given by the same formula as for lenses.

Magnification

The magnification produced by a mirror is also given by the same formula as for lenses.

Sample Working Example: Concave Mirror

Consider a concave mirror with a focal length of 15 cm. If an object is placed 30 cm from the mirror, we can use the mirror equation to calculate the image distance. The process is similar to the example for a converging lens.

Numerical Example: Convex Mirror Magnification

For a convex mirror with a magnification of 0.8, if the object height is 6 cm, we can find the image height using the magnification formula.

2.3 Optical Instruments

Optical instruments are devices designed to manipulate and enhance the observation or measurement of optical phenomena. They play a crucial role in various fields, including astronomy, microscopy, and photography.

2.3.1 The Human Eye

The human eye is a remarkable optical instrument that provides us with vision. Its main components include the cornea, lens, and retina. The lens is responsible for focusing light onto the retina, where the image is formed. The eye's accommodation ability allows it to adjust the focal length to focus on objects at different distances.

Lens Power

The power of the human eye's lens is determined by its focal length (f). The lens power (P) is given by the formula:

$$P = \frac{1}{f} \tag{2.14}$$

The unit of lens power is diopters (D).

Sample Working Example: Eye Prescription

If a person's prescription specifies a lens power of $+2.50$ D, the focal length of the corrective lens can be found using the formula for lens power.

Numerical Example: Near Point of Accommodation

The near point of accommodation (D_{near}) is the closest point at which the eye can focus. For a person with a lens power of $+3.00$ D, the near point can be calculated using the formula:

$$D_{near} = \frac{1}{P} \tag{2.15}$$

2.3.2 Microscopes

Microscopes are optical instruments that magnify small objects, enabling detailed observation. The compound microscope consists of an objective lens and an eyepiece. The total magnification is the product of the magnifications of the two lenses.

Total Magnification

The total magnification (M) of a compound microscope is given by:

$$M = M_{objective} \times M_{eyepiece} \tag{2.16}$$

Working Distance

The working distance is the distance between the object and the objective lens. It is related to the focal length of the objective lens.

Sample Working Example: Calculating Total Magnification

If an objective lens has a magnification of 40× and an eyepiece has a magnification of 10×, the total magnification of the compound microscope is 400×.

Numerical Example: Working Distance

For an objective lens with a focal length of 4 mm, the working distance can be calculated using the formula:

$$\text{Working Distance} = \frac{1}{\frac{1}{\text{Focal Length}} - \frac{1}{\text{Image Distance}}} \qquad (2.17)$$

2.3.3 Telescopes

Telescopes are optical instruments designed for observing distant objects in the sky. The two main types are refracting telescopes (using lenses) and reflecting telescopes (using mirrors).

Angular Magnification

The angular magnification (Mag) of a telescope is given by the formula:

$$\text{Mag} = \frac{\text{Focal Length of Telescope}}{\text{Focal Length of Eyepiece}} \qquad (2.18)$$

Resolving Power

The resolving power of a telescope determines its ability to distinguish fine details. It is proportional to the wavelength of light used and inversely proportional to the diameter of the telescope's aperture.

Sample Working Example: Angular Magnification

For a telescope with a focal length of 1000 mm and an eyepiece with a focal length of 10 mm, the angular magnification is 100×.

Numerical Example: Resolving Power

The resolving power (RP) of a telescope can be calculated using the formula:

$$\text{RP} = \frac{\lambda}{\text{Diameter of Aperture}} \tag{2.19}$$

Chapter 3

Wave Optics

3.1 Wave Nature of Light

Wave optics explores the phenomena associated with the wave nature of light, treating light as an electromagnetic wave. This perspective is essential for understanding interference, diffraction, and polarization.

3.1.1 Wave Equations

The wave nature of light is described by Maxwell's equations, a set of partial differential equations that govern the behavior of electric and magnetic fields in space. The general form of the wave equation for light is:

$$\frac{\partial^2 \mathbf{E}}{\partial t^2} = c^2 \nabla^2 \mathbf{E} \tag{3.1}$$

where $\mathbf{E}$ represents the electric field, c is the speed of light, and ∇^2 is the Laplacian operator.

3.1.2 Interference

Interference occurs when two or more coherent light waves superpose, leading to the reinforcement or cancellation of amplitudes. The intensity of the resulting wave is given by the formula:

$$I = I_1 + I_2 + 2\sqrt{I_1 I_2} \cos(\delta) \tag{3.2}$$

where I_1 and I_2 are the intensities of the individual waves, and δ is the phase difference between them.

Sample Working Example: Young's Double-Slit Experiment

In Young's double-slit experiment, a beam of light passes through two slits, creating an interference pattern on a screen. The fringe separation can be calculated using the formula:

$$\Delta y = \frac{\lambda L}{d} \tag{3.3}$$

where λ is the wavelength of light, L is the distance to the screen, and d is the slit separation.

Numerical Example: Interference Pattern

Consider a double-slit experiment with red light ($\lambda = 650$ nm), a screen 2 m away, and slits separated by 0.1 mm. The fringe separation is calculated as:

$$\Delta y = \frac{(650 \times 10^{-9} \text{ m})(2 \text{ m})}{0.1 \times 10^{-3} \text{ m}} \tag{3.4}$$

Thus, the interference pattern exhibits fringes approximately 13 mm apart.

3.1.3 Diffraction

Diffraction refers to the bending of light waves around obstacles or the spreading of light waves when passing through small openings. The intensity pattern in diffraction is given by the formula:

$$I(\theta) = I_0 \left(\frac{\sin(\alpha)}{\alpha} \right)^2 \tag{3.5}$$

where θ is the angle of observation, α is related to the aperture size and wavelength, and I_0 is the intensity at the center.

Sample Working Example: Single-Slit Diffraction

For a single slit of width a, the angular positions of minima in the diffraction pattern are given by:

$$a \sin(\theta) = m\lambda \tag{3.6}$$

where m is the order of the minimum.

Numerical Example: Single-Slit Diffraction Angle

Consider a single slit with a width of 0.02 mm and light with a wavelength of 500 nm. The angular position of the first minimum is calculated using the diffraction formula.

3.2 Interference and Diffraction

Interference and diffraction are fascinating wave phenomena that showcase the wave nature of light. Understanding these phenomena is crucial for explaining various optical patterns and behaviors.

3.2.1 Interference

Interference occurs when two or more waves overlap, leading to the reinforcement or cancellation of amplitudes. This phenomenon is fundamental to understanding optical patterns such as those seen in Young's double-slit experiment.

Young's Double-Slit Experiment

In this classic experiment, a beam of light passes through two closely spaced slits, creating an interference pattern on a screen. The pattern consists of alternating bright and dark fringes. The fringe separation (Δy) is determined by the wavelength of light (λ), the distance to the screen (L), and the slit separation (d):

$$\Delta y = \frac{\lambda L}{d} \tag{3.7}$$

This equation demonstrates the wave nature of light and the interference between the waves emanating from the two slits.

Sample Working Example: Calculating Fringe Separation

Consider a double-slit experiment with red light ($\lambda = 650$ nm), a screen 2 m away, and slits separated by 0.1 mm. The fringe separation can be calculated as:

$$\Delta y = \frac{(650 \times 10^{-9} \text{ m})(2 \text{ m})}{0.1 \times 10^{-3} \text{ m}} \tag{3.8}$$

This example illustrates the practical application of the interference formula.

3.2.2   Diffraction

Diffraction is the bending of light waves around obstacles or the spreading of light waves when passing through small openings. It is a phenomenon that becomes significant when the size of the aperture or obstacle is comparable to the wavelength of light.

Single-Slit Diffraction

For a single slit of width a, the angular positions of minima in the diffraction pattern are given by:

$$a \sin(\theta) = m\lambda \tag{3.9}$$

where m is the order of the minimum. This formula shows the dependence of diffraction patterns on the slit width and wavelength.

Sample Working Example: Calculating Diffraction Angle

Consider a single slit with a width of 0.02 mm and light with a wavelength of 500 nm. The angular position of the first minimum can be calculated using the diffraction formula.

Numerical Example: Diffraction Pattern Order

For a single slit of width 0.05 mm and green light ($\lambda = 550$ nm), if the observer is at an angle of 30° from the central maximum, the diffraction pattern order can be determined using:

$$a \sin(\theta) = m\lambda \tag{3.10}$$

This numerical example demonstrates the application of the single-slit diffraction formula.

3.3 Polarization

Polarization is a phenomenon associated with the orientation of oscillations in transverse waves. In optics, it is particularly relevant to describe the behavior of light waves and their interactions with materials.

3.3.1 Basics of Polarization

Light is an electromagnetic wave, and its electric field vector oscillates perpendicular to the direction of propagation. Polarization refers to the orientation of this electric field vector. When light waves undergo polarization, they are constrained to vibrate in specific planes.

Types of Polarization

There are various types of polarization, including linear polarization, circular polarization, and elliptical polarization. Linear polarization is common and involves the electric field oscillating in a straight line.

3.3.2 Polarizers and Malus's Law

Polarizers are optical devices that selectively allow light waves with a specific polarization to pass through. Malus's Law describes the intensity of light transmitted through a polarizer. If the incident light has intensity I_0 and makes an angle θ with the polarizer's transmission axis, the transmitted intensity I is given by:

$$I = I_0 \cos^2(\theta) \tag{3.11}$$

Sample Working Example: Polarizer Transmission

If unpolarized light with intensity 100 W/m^2 passes through a polarizer, and the polarizer's axis is at an angle of 45° with the incident light, the transmitted intensity can be calculated using Malus's Law.

Numerical Example: Polarization Efficiency

For a polarizer with an efficiency of 80%, if the incident light has an intensity of 200 W/m^2, the transmitted intensity can be determined using Malus's Law. This example demonstrates the practical application of polarization efficiency.

3.3.3   Birefringence and Wave Plates

Birefringence is a property of certain materials that exhibit two different refractive indices for light polarized in different directions. Wave plates, or retarders, are optical devices that exploit birefringence to alter the polarization state of light.

Retardance of Wave Plates

The retardance (δ) of a wave plate is related to its thickness and the birefringence of the material:

$$\delta = \frac{2\pi d}{\lambda}(n_e - n_o) \tag{3.12}$$

where d is the thickness, λ is the wavelength, and n_e and n_o are the extraordinary and ordinary refractive indices, respectively.

Sample Working Example: Wave Plate Retardance

For a wave plate with a thickness of 1 mm, a wavelength of 600 nm, and birefringence ($n_e - n_o$) of 0.1, the retardance can be calculated using the formula.

Numerical Example: Adjusting Polarization

If a linearly polarized light with an angle of polarization of 30° encounters a wave plate with a retardance of 90°, the output polarization state can be determined. This example illustrates the impact of a wave plate on the polarization of incident light.

Polarization plays a crucial role in understanding and manipulating light waves. It finds applications in technologies such as LCD displays, photography, and optical communications. The ability to control and analyze polarized light is fundamental in various optical systems.

Part II

Optical Systems

Chapter 4

Lens Systems

4.1 Thin Lens Formula

The thin lens formula is a fundamental equation in optics that relates the focal length of a lens to the object distance and image distance. It is particularly useful for understanding the behavior of lenses in various optical systems.

4.1.1 Thin Lens Equation

For a thin lens, the thin lens equation is given by:

$$\frac{1}{f} = \frac{1}{d_{\text{obj}}} + \frac{1}{d_{\text{img}}} \tag{4.1}$$

where f is the focal length of the lens, d_{obj} is the object distance, and d_{img} is the image distance.

Sign Convention

The sign convention for the thin lens equation is as follows: - f is positive for converging lenses and negative for diverging lenses. - d_{obj} is positive for objects on the same side as the incident light and negative for objects on the opposite side. - d_{img} is positive for images formed on the opposite side of the lens as the incident light and negative for images on the same side.

4.1.2   Magnification

The magnification (m) produced by a lens is given by the ratio of the image height (h_{img}) to the object height (h_{obj}):

$$m = \frac{h_{\text{img}}}{h_{\text{obj}}} \tag{4.2}$$

The magnification can also be expressed in terms of image distance and object distance:

$$m = -\frac{d_{\text{img}}}{d_{\text{obj}}} \tag{4.3}$$

Sample Working Example: Converging Lens

Consider a converging lens with a focal length of 10 cm. If an object is placed 20 cm from the lens, we can use the thin lens equation to calculate the image distance:

$$\frac{1}{f} = \frac{1}{d_{\text{obj}}} + \frac{1}{d_{\text{img}}} \tag{4.4}$$

Substituting values, we get:

$$\frac{1}{10 \text{ cm}} = \frac{1}{20 \text{ cm}} + \frac{1}{d_{\text{img}}} \tag{4.5}$$

Solving for d_{img}, we find $d_{\text{img}} = 20$ cm. Thus, the image is formed 20 cm from the lens.

Numerical Example: Magnification of Diverging Lens

For a diverging lens with a magnification of -0.5, if the object height is 4 cm, we can find the image height using the magnification formula:

$$m = \frac{h_{\text{img}}}{h_{\text{obj}}} \tag{4.6}$$

Substituting values, we get:

$$-0.5 = \frac{h_{\text{img}}}{4 \text{ cm}} \tag{4.7}$$

Solving for h_{img}, we find $h_{\text{img}} = -2$ cm. The negative sign indicates an inverted image.

4.1.3   Lens Systems

In optical systems involving multiple lenses, the combined focal length (F_{combined}) can be determined using the lensmaker's equation:

$$\frac{1}{F_{\text{combined}}} = \frac{1}{f_1} + \frac{1}{f_2} - \frac{d}{f_1 f_2} \tag{4.8}$$

where f_1 and f_2 are the focal lengths of the individual lenses, and d is the distance between the lenses.

Sample Working Example: Lens System

Consider two lenses with focal lengths $f_1 = 15$ cm and $f_2 = -20$ cm placed 30 cm apart. Using the lensmaker's equation, we can find the combined focal length of the system.

Numerical Example: Image Formation in Lens System

For a lens system with two converging lenses ($f_1 = 10$ cm, $f_2 = 15$ cm) separated by 25 cm, if an object is placed 30 cm from the first lens, we can use the thin lens formula to calculate the final image distance.

4.2   Lens Aberrations

Lens aberrations are deviations from ideal optical behavior that can affect the quality of images formed by lenses. Understanding these aberrations is crucial for designing optical systems that produce clear and sharp images.

4.2.1   Types of Lens Aberrations

There are several types of lens aberrations, each with its characteristics and impact on image quality. Some common lens aberrations include:

Chromatic Aberration

Chromatic aberration occurs when different wavelengths of light are refracted by different amounts, leading to color fringing around the edges of objects. It can be minimized using achromatic lenses that combine different materials to reduce the dispersion effect.

Spherical Aberration

Spherical aberration results from the spherical shape of lenses, causing parallel rays of light to converge at different points, leading to blurred images. This can be reduced by using aspherical lens surfaces.

Coma

Coma aberration causes off-axis light rays to produce comet-shaped aberrations, especially in wide-angle optical systems. It can be mitigated through careful lens design and choosing appropriate lens shapes.

Astigmatism

Astigmatism occurs when different meridians of the lens have different focal lengths, leading to distorted images. Corrective lenses with cylindrical surfaces are used to address astigmatism.

4.2.2   Aberration Correction

Lens aberrations can be corrected using various techniques, including combining multiple lenses, using specialized lens shapes, and employing corrective elements.

Apochromatic Lenses

Apochromatic lenses are designed to minimize chromatic aberration by combining multiple lens elements made of different materials to bring different wavelengths of light to a common focus.

Use of Aspherical Surfaces

Aspherical lens surfaces are designed to reduce spherical aberration by deviating from a perfect spherical shape. They are often used in high-performance optical systems.

4.2.3 Sample Working Example: Chromatic Aberration Correction

Consider a lens with significant chromatic aberration, producing color fringes around the image. To correct this, an apochromatic lens can be used, consisting of multiple lens elements made of different glasses. This example illustrates the practical application of aberration correction.

4.2.4 Numerical Example: Spherical Aberration Reduction

For a lens system with noticeable spherical aberration, the use of an aspherical lens surface can be explored. The aspherical surface helps bring parallel rays of light to a common focus point, reducing the effects of spherical aberration.

4.3 Lens Combinations

Lens combinations involve the arrangement of multiple lenses in optical systems, and understanding their behavior is crucial for designing complex optical setups. The combination of lenses introduces new parameters and characteristics that impact the overall performance of the system.

4.3.1 Lens Combination Formula

The lens combination formula relates the focal length of a lens combination (F_{combined}) to the focal lengths of individual lenses (f_1 and f_2) and the separation between the lenses (d):

$$\frac{1}{F_{\text{combined}}} = \frac{1}{f_1} + \frac{1}{f_2} - \frac{d}{f_1 f_2} \qquad (4.9)$$

This formula is derived from the lensmaker's equation and provides insights into the behavior of lens systems.

4.3.2   Magnification in Lens Systems

The magnification (m_{combined}) of a lens system can be calculated using the magnification of individual lenses and the separation between them. For two lenses separated by a distance D, the combined magnification is given by:

$$m_{\text{combined}} = m_1 \times m_2 \qquad (4.10)$$

where m_1 and m_2 are the magnifications of the individual lenses.

Sample Working Example: Lens Combination Magnification

Consider two lenses with magnifications $m_1 = -2$ and $m_2 = 3$ separated by a distance $D = 15$ cm. The combined magnification can be calculated using the formula, providing insights into the overall image size.

4.3.3   Aberrations in Lens Systems

Lens combinations can introduce or mitigate aberrations depending on the arrangement of lenses. Understanding the types of aberrations and their impact on the final image is essential for optimizing optical systems.

Minimizing Chromatic Aberration

To minimize chromatic aberration in a lens system, using an achromatic doublet (a combination of lenses made of different materials) can be effective. This helps bring different wavelengths of light to a common focus.

Addressing Spherical Aberration

Spherical aberration can be reduced by carefully selecting the curvature of lens surfaces and using aspherical lenses. The arrangement of lenses in a system plays a crucial role in managing aberrations.

4.3.4 Sample Working Example: Lens System Design

Consider designing a lens system for a camera with specific requirements, such as minimizing chromatic aberration and maximizing magnification. The combination of lenses, their focal lengths, and separations can be optimized to meet these criteria.

4.3.5 Numerical Example: Lens System Optimization

For a telescope requiring a high magnification of 50x and minimal chromatic aberration, a lens system is designed using an achromatic doublet. The focal lengths and separation are chosen to achieve the desired performance.

Chapter 5

Mirrors and Reflective Systems

5.1 Mirror Equations

Mirror equations are fundamental in understanding the behavior of mirrors and reflective systems. These equations describe the relationships between object distance, image distance, and focal length for both concave and convex mirrors.

5.1.1 Concave Mirror Equations

For a concave mirror, the mirror equation relates the object distance (d_{obj}), image distance (d_{img}), and focal length (f):

$$\frac{1}{f} = \frac{1}{d_{\text{obj}}} + \frac{1}{d_{\text{img}}} \tag{5.1}$$

This equation is crucial for determining where an image is formed and whether it is real or virtual.

Magnification in Concave Mirrors

The magnification (m_{concave}) produced by a concave mirror is given by the ratio of image height (h_{img}) to object height (h_{obj}):

$$\begin{aligned} m_{\text{concave}} &= -\frac{d_{\text{img}}}{d_{\text{obj}}} \\ &= \frac{h_{\text{img}}}{h_{\text{obj}}} \end{aligned} \tag{5.2}$$

The negative sign indicates that the image is inverted.

5.1.2   Convex Mirror Equations

For a convex mirror, the mirror equation is modified to account for the virtual focus and negative focal length:

$$\frac{1}{f} = \frac{1}{d_{\text{obj}}} - \frac{1}{d_{\text{img}}} \tag{5.3}$$

This equation helps determine the location of the virtual image formed by a convex mirror.

Magnification in Convex Mirrors

The magnification (m_{convex}) for a convex mirror is also given by the ratio of image height to object height:

$$\begin{aligned} m_{\text{convex}} &= \frac{d_{\text{img}}}{d_{\text{obj}}} \\ &= -\frac{h_{\text{img}}}{h_{\text{obj}}} \end{aligned} \tag{5.4}$$

In convex mirrors, the image is always virtual and upright.

5.1.3   Sample Working Example: Concave Mirror Image Formation

Consider a concave mirror with a focal length of 20 cm. If an object is placed 30 cm away, we can use the mirror equation to calculate the image distance and magnification.

5.1.4 Numerical Example: Convex Mirror Virtual Image

For a convex mirror with a focal length of 15 cm, if an object is located 10 cm in front of the mirror, we can use the mirror equation to find the image distance and magnification. This example demonstrates the characteristics of images formed by convex mirrors.

5.2 Types of Mirrors

Mirrors play a crucial role in optical systems, and different types of mirrors serve various purposes in shaping and redirecting light. Understanding the characteristics of concave and convex mirrors is essential for designing reflective systems with specific optical outcomes.

5.2.1 Concave Mirrors

Concave mirrors are curved inward, and they have unique optical properties. The mirror equation for concave mirrors is given by:

$$\frac{1}{f} = \frac{1}{d_{\mathrm{obj}}} + \frac{1}{d_{\mathrm{img}}} \tag{5.5}$$

where f is the focal length, d_{obj} is the object distance, and d_{img} is the image distance.

Real and Virtual Images

Concave mirrors can form both real and virtual images. Real images are formed when the reflected light converges, whereas virtual images are formed when the reflected rays appear to diverge.

Magnification

The magnification (m_{concave}) produced by a concave mirror is given by:

$$
\begin{aligned}
m_{\text{concave}} &= -\frac{d_{\text{img}}}{d_{\text{obj}}} \\
&= \frac{h_{\text{img}}}{h_{\text{obj}}}
\end{aligned}
\tag{5.6}
$$

The negative sign indicates inversion.

5.2.2   Convex Mirrors

Convex mirrors are curved outward, and they have distinct reflective properties. The mirror equation for convex mirrors is modified to:

$$
\frac{1}{f} = \frac{1}{d_{\text{obj}}} - \frac{1}{d_{\text{img}}}
\tag{5.7}
$$

Virtual Images in Convex Mirrors

Convex mirrors always produce virtual images, regardless of the object's position. These virtual images are upright and diminished.

Magnification in Convex Mirrors

The magnification (m_{convex}) for a convex mirror is given by:

$$
\begin{aligned}
m_{\text{convex}} &= \frac{d_{\text{img}}}{d_{\text{obj}}} \\
&= -\frac{h_{\text{img}}}{h_{\text{obj}}}
\end{aligned}
\tag{5.8}
$$

5.2.3   Sample Working Example: Concave Mirror Imaging

Consider a concave mirror with a focal length of 30 cm. If an object is placed 40 cm away, calculate the image distance and magnification using the concave mirror equation.

5.2.4 Numerical Example: Convex Mirror Image Formation

For a convex mirror with a focal length of 25 cm, if an object is located 15 cm in front of the mirror, determine the image distance and magnification using the convex mirror equation.

5.3 Reflective Systems Design

Designing reflective optical systems involves the careful selection and arrangement of mirrors to achieve specific optical outcomes. The interplay between mirror types, their positions, and distances contributes to the overall functionality of reflective systems.

5.3.1 Mirror Combination for Imaging

In designing a reflective system for imaging, the combination of concave and convex mirrors can be utilized. The mirror combination formula is crucial for determining the combined focal length ($F_{combined}$) of the system:

$$\frac{1}{F_{combined}} = \frac{1}{f_{concave}} + \frac{1}{f_{convex}} \tag{5.9}$$

where $f_{concave}$ and f_{convex} are the focal lengths of the concave and convex mirrors, respectively.

Sample Working Example: Mirror Combination for Imaging

Consider a reflective system consisting of a concave mirror with a focal length of 20 cm and a convex mirror with a focal length of 15 cm. Calculate the combined focal length of the system using the mirror combination formula.

Numerical Example: Optimizing Imaging System

For a telescope design requiring a specific magnification, the reflective system can be optimized by choosing appropriate concave and convex mirrors. The mirror combination formula aids in determining the necessary focal lengths to achieve the desired magnification.

5.3.2   Mirror Systems for Light Collection

Reflective systems are often used in astronomical instruments to collect and focus light. The light collection efficiency of a reflective system can be calculated using the effective aperture diameter (D_{eff}) and the focal length (f):

$$D_{\text{eff}} = 2 \cdot f \cdot \tan\left(\frac{\theta}{2}\right) \tag{5.10}$$

where θ is the angular field of view.

Sample Working Example: Light Collection Efficiency

For a reflective system with a focal length of 50 cm and an angular field of view of 0.02 radians, calculate the effective aperture diameter using the light collection formula.

Numerical Example: Telescope Design for Light Collection

In designing a telescope for optimal light collection, the choice of mirrors and their arrangement plays a critical role. The light collection formula assists in determining the effective aperture diameter required for a specific field of view.

Part III

Light Sources

Chapter 6

Light Emission

6.1 Incandescent Sources

Incandescent sources are a type of light-emitting device that produces illumination through the heating of a material to high temperatures, causing it to emit visible light. Understanding the principles and characteristics of incandescent sources is essential in the study of light emission.

6.1.1 Blackbody Radiation

The emission of light from incandescent sources is governed by blackbody radiation. The spectral radiance (B_λ) of a blackbody is described by Planck's law:

$$B_\lambda(\lambda, T) = \frac{8\pi hc}{\lambda^5} \cdot \frac{1}{e^{\frac{hc}{\lambda kT}} - 1} \tag{6.1}$$

where λ is the wavelength, T is the temperature, h is Planck's constant, c is the speed of light, and k is the Boltzmann constant.

6.1.2   Incandescent Lamp Efficiency

The efficiency of an incandescent lamp in converting electrical power to visible light is given by the luminous efficacy (η):

$$\eta = \frac{\text{Visible Light Output (lumens)}}{\text{Electrical Power Input (watts)}} \tag{6.2}$$

Sample Working Example: Incandescent Lamp Efficiency

Consider an incandescent lamp that emits 800 lumens of visible light and consumes 60 watts of electrical power. Calculate the luminous efficacy of the lamp.

6.1.3   Color Temperature of Incandescent Sources

The color temperature (T_{color}) of incandescent sources is related to the perceived color of the emitted light. It is calculated using Wien's displacement law:

$$T_{\text{color}} = \frac{b}{\lambda_{\text{max}}} \tag{6.3}$$

where b is Wien's displacement constant and λ_{max} is the wavelength at which the spectral radiance is maximized.

Numerical Example: Determining Color Temperature

For an incandescent source with a spectral radiance peak at 600 nm, calculate the color temperature using Wien's displacement law.

6.1.4   Incandescent Sources in Lighting Design

Incandescent sources have been widely used in lighting design, but their inefficiency and low lifespan have led to the development of more energy-efficient alternatives. Understanding the principles of incandescent sources remains valuable in historical and educational contexts.

Impact of Incandescent Sources on Color Rendering

Incandescent sources are known for their high color rendering index (CRI), making them suitable for applications where accurate color representation is essential.

Historical Significance

Incandescent lamps have a rich history and played a significant role in early lighting technologies. Exploring their principles provides insights into the evolution of lighting.

6.2   Gas Discharge Sources

Gas discharge sources are a class of light-emitting devices that produce illumination through the excitation of gases or vapors within a discharge tube. Understanding the principles and characteristics of gas discharge sources is essential in the study of light emission.

6.2.1   Spectral Emission Lines

Gas discharge sources emit light at specific wavelengths corresponding to the spectral lines of the excited gas or vapor. The spectral radiance (I_λ) of a gas discharge source can be described by:

$$I_\lambda = A \cdot P \cdot f(\lambda) \tag{6.4}$$

where A is the cross-sectional area of the discharge tube, P is the pressure of the gas, and $f(\lambda)$ represents the normalized spectral emission function.

6.2.2   Gas Discharge Lamp Efficiency

The efficiency of a gas discharge lamp in converting electrical power to visible light is given by the luminous efficacy (η):

$$\eta = \frac{\text{Visible Light Output (lumens)}}{\text{Electrical Power Input (watts)}} \tag{6.5}$$

Sample Working Example: Gas Discharge Lamp Efficiency

Consider a gas discharge lamp that emits 2000 lumens of visible light and consumes 30 watts of electrical power. Calculate the luminous efficacy of the lamp.

6.2.3   Color Rendering Index (CRI) of Gas Discharge Lamps

The color rendering index is a measure of a light source's ability to accurately render colors compared to a reference light source. Gas discharge lamps typically have a high CRI, making them suitable for applications where color fidelity is crucial.

Numerical Example: Calculating CRI

For a gas discharge lamp with known spectral emission lines, calculate the color rendering index using the standard method.

6.2.4   Gas Discharge Sources in Practical Applications

Gas discharge sources find widespread use in various practical applications, including street lighting, fluorescent lighting, and specialized lighting in scientific and industrial settings.

Fluorescent Lamps

Fluorescent lamps are a common type of gas discharge source that produces visible light through the fluorescence of phosphor coatings inside the lamp.

High-Intensity Discharge (HID) Lamps

HID lamps, such as metal halide and sodium vapor lamps, are employed for high-intensity lighting in outdoor and industrial environments.

Plasma Displays

Gas discharge principles are utilized in plasma display panels (PDPs), which have been historically used in television screens and monitors.

6.3 Semiconductor Sources

Semiconductor sources, including light-emitting diodes (LEDs) and laser diodes, are key components in modern lighting and communication technologies. Understanding the principles and characteristics of semiconductor sources is crucial in the study of light emission.

6.3.1 LED Emission Spectrum

Light-emitting diodes (LEDs) emit light across a specific spectrum determined by the semiconductor material used. The radiance spectrum (L_λ) of an LED is characterized by:

$$L_\lambda(\lambda) = \frac{I_\lambda(\lambda)}{\pi r^2} \tag{6.6}$$

where $I_\lambda(\lambda)$ is the radiant intensity and r is the distance from the LED.

6.3.2 LED Efficiency and Luminous Flux

The efficiency of an LED in converting electrical power to visible light is given by the luminous efficacy (η):

$$\eta = \frac{\text{Luminous Flux (lumens)}}{\text{Electrical Power Input (watts)}} \tag{6.7}$$

Sample Working Example: LED Efficiency

Consider an LED that emits 500 lumens of visible light and consumes 5 watts of electrical power. Calculate the luminous efficacy of the LED.

6.3.3   Laser Diode Characteristics

Laser diodes are semiconductor devices that emit coherent and monochromatic light. The output power (P_{out}) of a laser diode can be related to the injected current (I_{inj}) by the diode's efficiency factor (η_{diode}):

$$P_{out} = \eta_{diode} \cdot I_{inj} \tag{6.8}$$

Numerical Example: Laser Diode Output Power

For a laser diode with an efficiency factor of 0.8, if the injected current is 100 mA, calculate the output power of the laser diode.

6.3.4   Applications of Semiconductor Sources

Semiconductor sources have found widespread applications in various fields, ranging from lighting and displays to telecommunications and medical technologies.

LEDs in Illumination

Light-emitting diodes are extensively used in illumination applications due to their energy efficiency, long lifespan, and compact size.

Laser Diodes in Optical Communications

Laser diodes play a crucial role in optical communication systems, where their coherent light emission enables high-speed data transmission through optical fibers.

Medical Imaging with LEDs

In medical imaging, LEDs are employed for applications such as endoscopy and phototherapy due to their flexibility and controllable spectral characteristics.

Chapter 7

Lasers and Coherent Sources

7.1 Laser Principles

Lasers (Light Amplification by Stimulated Emission of Radiation) are coherent light sources that have found extensive applications in various fields, including communications, medical procedures, and material processing. Understanding the principles underlying laser operation is essential for those working in the field of optics.

7.1.1 Population Inversion

Laser operation relies on achieving a population inversion, where more atoms or molecules are in higher energy states than in lower energy states. This inversion is typically achieved through pumping mechanisms, such as optical or electrical excitation.

7.1.2 Stimulated Emission

Stimulated emission is a process crucial to laser operation. When an atom or molecule in an excited state encounters a photon of the same energy, it can undergo stimulated emission, releasing a second photon with the same energy and phase.

7.1.3 Gain Medium and Resonator Cavity

The active medium in a laser, known as the gain medium, amplifies light through stimulated emission. The gain medium is placed within a resonator cavity, which consists of two mirrors—one highly reflective and the other partially transparent. The resonator cavity helps maintain the stimulated emission process and provides directionality to the emitted light.

7.1.4 Conditions for Laser Action

For laser action to occur, three conditions must be satisfied: population inversion, stimulated emission dominance over absorption, and feedback through the resonator cavity to create coherent and collimated light.

7.1.5 Laser Output Characteristics

The characteristics of laser output are influenced by factors such as the gain medium, resonator design, and pumping mechanism. Laser output is typically characterized by its wavelength, coherence, and directionality.

Sample Working Example: Calculating Laser Wavelength

Consider a laser with a gain medium that emits light with a wavelength of 632.8 nm. Calculate the energy associated with this wavelength using Planck's equation.

Numerical Example: Laser Coherence Length

For a helium-neon laser with a wavelength of 632.8 nm, calculate the coherence length using the formula $L_{\text{coherence}} = \frac{\lambda^2}{\Delta\lambda}$, where $\Delta\lambda$ is the spectral linewidth.

7.1.6 Applications of Lasers

Lasers have diverse applications in various fields, ranging from telecommunications to medical procedures and scientific research.

Telecommunications

Lasers are widely used in optical fiber communication systems for high-speed data transmission.

Medical Applications

In medicine, lasers are employed for surgeries, diagnostics, and treatments due to their precision and ability to interact with biological tissues.

Material Processing

Lasers are used for cutting, welding, and engraving in industries, leveraging their focused and intense beams.

7.2 Types of Lasers

Lasers come in various types, each tailored for specific applications based on their unique characteristics and operating principles. Understanding the different types of lasers is essential for engineers and researchers working with laser technology.

7.2.1 Gas Lasers

Gas lasers use a gaseous medium as the active gain medium. Common examples include helium-neon (HeNe) lasers and carbon dioxide (CO2) lasers. Gas lasers cover a broad range of wavelengths, making them versatile for applications such as spectroscopy and laser cutting.

Example: Helium-Neon (HeNe) Laser

The helium-neon laser operates in the visible spectrum, commonly emitting red light at a wavelength of 632.8 nm. It is widely used in alignment, interferometry, and educational demonstrations.

7.2.2 Solid-State Lasers

Solid-state lasers utilize a solid gain medium, typically a crystal or glass doped with rare-earth ions. Examples include neodymium-doped yttrium aluminum garnet (Nd:YAG) lasers and erbium-doped fiber lasers. Solid-state lasers find applications in material processing, medical procedures, and telecommunications.

Example: Nd:YAG Laser

The neodymium-doped yttrium aluminum garnet (Nd:YAG) laser emits infrared light at a wavelength of 1064 nm. It is used in industrial cutting and welding, as well as medical applications.

7.2.3 Semiconductor Lasers

Semiconductor lasers, also known as diode lasers, use a semiconductor material as the gain medium. They are compact, efficient, and widely used in telecommunications, optical storage devices, and laser pointers.

Example: Semiconductor Laser Diode

A semiconductor laser diode emits coherent light through stimulated emission in a semiconductor junction. Laser diodes are commonly found in optical communication systems, barcode scanners, and laser printers.

7.2.4 Fiber Lasers

Fiber lasers use an optical fiber as the gain medium. They offer high power, excellent beam quality, and are employed in materials processing, laser surgery, and military applications.

Example: Ytterbium-Doped Fiber Laser

Ytterbium-doped fiber lasers operate in the infrared region and are known for their high efficiency. They are used in industrial applications such as cutting, welding, and marking.

7.2.5 Free-Electron Lasers

Free-electron lasers (FELs) utilize accelerated electrons as the gain medium. They provide tunable and powerful laser beams and are used in scientific research, medical imaging, and defense applications.

Example: Stanford Linear Collider (SLAC)

The Stanford Linear Collider is an example of a free-electron laser facility, producing intense X-ray beams for advanced scientific experiments.

7.3 Applications of Lasers

Lasers have revolutionized various industries and scientific fields with their unique properties and capabilities. Understanding the diverse applications of lasers is crucial for engineers, researchers, and practitioners in the field of optics.

7.3.1 Industrial Applications

Lasers find extensive use in industrial processes due to their precision and ability to generate focused and intense beams.

Laser Cutting and Welding

High-power lasers, such as CO2 and fiber lasers, are employed for cutting and welding metals and non-metallic materials in industries such as automotive and aerospace.

Marking and Engraving

Lasers are used for marking and engraving products, providing permanent and high-quality markings on materials like plastics, metals, and ceramics.

Additive Manufacturing

In additive manufacturing, lasers are used in processes like selective laser sintering (SLS) and stereolithography (SLA) to build three-dimensional objects layer by layer.

7.3.2　Medical Applications

Lasers have transformed the field of medicine, offering minimally invasive procedures and precise treatments.

Laser Surgery

Lasers are used in various surgical procedures, including eye surgeries (LASIK), dermatological procedures, and dental surgeries, providing high precision and reduced recovery times.

Photodynamic Therapy

Photodynamic therapy (PDT) uses lasers to activate photosensitive drugs, selectively destroying cancer cells in treatments for certain types of cancer.

Diagnostic Imaging

Lasers are employed in diagnostic imaging techniques such as laser-induced fluorescence and optical coherence tomography (OCT) for non-invasive imaging of tissues.

7.3.3 Communications and Information Technology

Lasers play a crucial role in optical communication and information technology.

Fiber Optic Communication

Semiconductor lasers are used in fiber optic communication systems for transmitting data over long distances with high bandwidth and minimal signal loss.

Optical Data Storage

Laser diodes are integral components in optical data storage devices such as CD, DVD, and Blu-ray systems, facilitating high-capacity storage.

Lidar Technology

Lidar (Light Detection and Ranging) systems utilize lasers to measure distances and create detailed, three-dimensional maps, finding applications in autonomous vehicles and environmental monitoring.

7.3.4 Scientific Research

Lasers are indispensable tools in scientific research across various disciplines.

Spectroscopy

Lasers are used in spectroscopic techniques to study the interaction of light with matter, providing valuable information about molecular structures and chemical processes.

Ultrafast Laser Physics

Ultrafast lasers, with pulse durations in the femtosecond range, enable studies in ultrafast physics, attosecond science, and high-precision measurements.

Nuclear Fusion Research

High-power lasers, like those used in inertial confinement fusion experiments, contribute to research in controlled nuclear fusion for energy production.

Part IV

Optics in Practice

Chapter 8

Optical Materials

8.1 Optical Properties of Materials

The optical properties of materials play a crucial role in the behavior of light as it interacts with substances. Understanding these properties is fundamental for designing optical systems and analyzing the transmission, reflection, and absorption of light.

8.1.1 Index of Refraction

The index of refraction (n) is a key optical property that describes how light propagates through a medium. It is defined as the ratio of the speed of light in a vacuum (c) to the speed of light in the material (v):

$$n = \frac{c}{v} \tag{8.1}$$

Example: Snell's Law

Snell's Law relates the angles of incidence (θ_1) and refraction (θ_2) to the indices of refraction of two media:

$$n_1 \sin(\theta_1) = n_2 \sin(\theta_2) \tag{8.2}$$

8.1.2 Dispersion

Dispersion refers to the variation of the index of refraction with the wavelength of light. It causes different colors to experience different degrees of refraction.

Example: Chromatic Aberration

Chromatic aberration in lenses occurs due to dispersion, causing different colors to focus at different points. This can be corrected using achromatic lenses.

8.1.3 Absorption and Transmission

Materials can absorb and transmit light to varying degrees. The absorption coefficient (α) quantifies the fraction of incident light absorbed per unit distance.

Example: Beer-Lambert Law

The Beer-Lambert Law expresses the relationship between absorption, concentration (C), and the path length (l):

$$A = \log\left(\frac{I_0}{I}\right) = \varepsilon \cdot C \cdot l \tag{8.3}$$

where A is the absorbance, I_0 is the incident intensity, I is the transmitted intensity, and ε is the molar absorptivity.

8.1.4 Reflection and Transmission Coefficients

The reflection coefficient (R) and transmission coefficient (T) describe the amount of light reflected and transmitted at an interface.

Example: Fresnel Equations

The Fresnel equations provide a mathematical description of the reflection and transmission coefficients for light incident on a dielectric interface.

8.1.5 Optical Birefringence

Birefringence is an optical property where a material exhibits different refractive indices for light polarized in different directions.

Example: Birefringent Materials

Calcite is an example of a birefringent material, commonly used in polarizing prisms and optical devices.

8.2 Selection of Optical Materials

The selection of optical materials is a critical aspect of designing optical systems, influencing factors such as transmission, dispersion, and mechanical stability. Engineers must carefully choose materials based on the specific requirements of their application.

8.2.1 Transparency and Absorption

The transparency of a material to certain wavelengths of light is crucial for applications such as lenses and windows. Consider the absorption characteristics of materials to ensure minimal loss of light.

Example: UV-Visible Spectral Range

For applications requiring transparency in the UV-visible range, materials like fused silica and optical glasses are commonly selected due to their low absorption in this spectral region.

8.2.2 Index of Refraction Matching

Matching the indices of refraction between different optical elements reduces reflection and improves the efficiency of optical systems.

Example: Anti-Reflective Coatings

Anti-reflective coatings are applied to optical surfaces to minimize reflections by adjusting the refractive index to match that of the surrounding medium.

8.2.3   Thermal Stability

Optical materials must maintain their properties under varying temperature conditions, especially in applications where temperature fluctuations are significant.

Example: Infrared Imaging Systems

Materials like germanium and zinc selenide are chosen for infrared imaging systems due to their thermal stability and transparency in the infrared spectrum.

8.2.4   Mechanical Strength

Consider the mechanical strength and durability of materials, especially in applications where optical elements may be subjected to mechanical stress or harsh environments.

Example: Lens Mounts

For lens mounts in rugged environments, materials like aluminum alloys or titanium may be preferred for their strength and lightweight properties.

8.2.5   Cost Considerations

The cost of materials is a practical consideration, especially in large-scale production or when designing cost-sensitive optical systems.

Example: Consumer Optics

In consumer optics, materials like acrylic or polycarbonate may be chosen for lenses and displays due to their cost-effectiveness and acceptable optical properties.

8.2.6 Environmental Compatibility

Consider the compatibility of optical materials with the environment in which they will be used, especially in applications exposed to corrosive substances or extreme conditions.

Example: Space Optics

Optical materials for space applications must withstand the harsh conditions of space, including vacuum, radiation, and temperature extremes.

8.2.7 Custom Material Requirements

In some cases, specific material properties may be required for unique optical applications, leading to the development or customization of optical materials.

Example: Nonlinear Optical Materials

Materials with specific nonlinear optical properties are crucial in applications like frequency conversion or parametric amplification in lasers.

Chapter 9

Optical Coatings

9.1 Thin Film Coatings

Thin film coatings play a pivotal role in optics, providing control over the reflection, transmission, and absorption of light. Understanding the principles behind thin film coatings is essential for designing optical systems with specific performance characteristics.

9.1.1 Optical Interference

Thin film coatings rely on the principles of optical interference to enhance or suppress certain wavelengths of light.

Example: Anti-Reflective Coating

An anti-reflective coating is designed to minimize reflection by creating constructive interference for certain wavelengths. The coating thickness is tailored to achieve destructive interference for the desired wavelength range.

9.1.2   Film Thickness and Refractive Index

The thickness and refractive index of the thin film coatings directly impact their optical properties.

Example: Fabry-Perot Interferometer

A Fabry-Perot interferometer utilizes multiple thin film layers with varying thickness to create interference patterns, enabling precise measurements of wavelengths and optical properties.

9.1.3   Coating Materials

Different materials are used for thin film coatings depending on the desired optical characteristics and application requirements.

Example: Dielectric Coatings

Dielectric coatings, composed of non-conductive materials like metal oxides, are commonly used for interference coatings due to their high transparency and customizable refractive indices.

9.1.4   Multilayer Coatings

Multilayer coatings consist of multiple thin film layers stacked together, allowing for more complex control over optical properties.

Example: Bandpass Filters

Bandpass filters use multiple layers of thin films to transmit light within a specific wavelength range while blocking others, finding applications in optical communication and spectroscopy.

9.1.5   Thickness Monitoring and Control

The accurate monitoring and control of thin film thickness during deposition are critical for achieving the desired optical performance.

Example: Optical Monitoring Systems

Optical monitoring systems, such as quartz crystal monitors, provide real-time feedback during deposition, allowing for precise control of film thickness.

9.1.6   Durability and Environmental Stability

Thin film coatings must be designed to withstand environmental conditions and maintain their optical properties over time.

Example: Hard Coatings

Hard coatings, often composed of materials like metal nitrides, enhance the durability and scratch resistance of optical surfaces.

9.1.7   Applications of Thin Film Coatings

Thin film coatings find widespread applications in various optical devices and systems.

Example: Camera Lenses

Camera lenses often incorporate anti-reflective coatings to improve light transmission and reduce unwanted reflections, enhancing image quality.

9.2   Anti-reflective Coatings

Anti-reflective coatings are essential in optics to minimize unwanted reflections and enhance the transmission of light through optical surfaces. Understanding the principles behind anti-reflective coatings is crucial for designing optical systems with improved performance.

9.2.1   Principle of Anti-Reflection

Anti-reflective coatings work on the principle of interference, aiming to reduce reflections by creating destructive interference for specific wavelengths.

Example: Single-layer Coating

A single-layer anti-reflective coating can be designed to have a thickness that results in destructive interference for a certain wavelength, minimizing reflection.

9.2.2   Design Considerations

The design of anti-reflective coatings involves considerations such as the desired wavelength range, incident angle, and refractive indices of the coating and substrate.

Example: Broadband Anti-reflective Coating

A broadband anti-reflective coating may consist of multiple layers with varying thickness, allowing for effective suppression of reflections across a wide range of wavelengths.

9.2.3   Reflectance and Transmittance

The effectiveness of an anti-reflective coating is often quantified in terms of reflectance (R) and transmittance (T).

Example: Reflectance Calculation

The reflectance of an optical surface with an anti-reflective coating can be calculated using the formula:

$$R = \left(\frac{n_2 - n_1}{n_2 + n_1} \right)^2 \tag{9.1}$$

where n_1 is the refractive index of the coating, and n_2 is the refractive index of the substrate.

9.2.4 Double-layer Anti-reflective Coatings

Double-layer anti-reflective coatings utilize two layers with different refractive indices to achieve enhanced performance.

Example: MgF2/SiO2 Coating

A common double-layer coating consists of magnesium fluoride (MgF_2) and silicon dioxide (SiO_2), providing excellent anti-reflective properties in the visible spectrum.

9.2.5 Multiple-layer Anti-reflective Coatings

Multiple-layer coatings further optimize anti-reflective performance by incorporating additional layers.

Example: Quarter-wave Stack

A quarter-wave stack consists of alternating high and low refractive index layers, achieving broad-spectrum anti-reflective properties.

9.2.6 Application in Eyeglasses

Anti-reflective coatings are widely used in eyeglasses to improve vision by reducing glare and reflections.

Example: Eyeglass Coating

An anti-reflective coating on eyeglasses enhances visual comfort by allowing more light to pass through the lenses, reducing reflections from both the front and back surfaces.

9.2.7 Durability and Maintenance

Considerations for the durability and ease of maintenance are crucial in practical applications of anti-reflective coatings.

Example: Hard Coatings

Hard coatings, often applied to anti-reflective coatings, enhance resistance to scratches and abrasions, ensuring longevity.

9.3   Coating Techniques

The application of optical coatings involves various techniques to deposit thin films onto surfaces, influencing the performance and characteristics of optical components. Understanding different coating techniques is crucial for optimizing the properties of optical coatings.

9.3.1   Evaporation Deposition

Evaporation deposition is a common technique where coating materials are heated in a vacuum to generate vapor that condenses on the substrate.

Example: Metal Mirrors

Metal mirrors often use aluminum deposited through evaporation, providing a reflective coating for optical surfaces.

9.3.2   Sputter Deposition

Sputter deposition involves bombarding a target material with energetic ions to release atoms, forming a thin film on the substrate.

Example: Dielectric Coatings

Dielectric coatings, such as those used in interference filters, are often deposited using sputter deposition for precise control over film properties.

9.3.3 Chemical Vapor Deposition (CVD)

Chemical Vapor Deposition utilizes chemical reactions to deposit thin films on a substrate, offering high uniformity and conformal coating.

Example: Semiconductor Devices

CVD is commonly used in the production of semiconductor devices, providing thin films for various electronic components.

9.3.4 Atomic Layer Deposition (ALD)

Atomic Layer Deposition is a precise technique where thin films are built layer by layer, enabling accurate control over film thickness.

Example: Nanotechnology Applications

ALD is essential in nanotechnology for depositing thin films with nanoscale precision, influencing properties at the atomic level.

9.3.5 Dip Coating

Dip coating involves immersing a substrate into a solution containing the coating material and withdrawing it at a controlled rate.

Example: Protective Coatings

Dip coating is used for applying protective coatings on lenses or optical elements, enhancing durability and resistance to environmental factors.

9.3.6 Spin Coating

Spin coating utilizes centrifugal force to spread a liquid coating material evenly over a rotating substrate, producing uniform thin films.

Example: Thin Photoresist Layers

In microfabrication processes, spin coating is employed to apply thin and uniform layers of photoresist for lithography.

9.3.7   Plasma Enhanced Chemical Vapor Deposition (PECVD)

PECVD is a variation of CVD that uses plasma to enhance chemical reactions, allowing for lower deposition temperatures and improved film quality.

Example: Thin Film Solar Cells

PECVD is widely used in the production of thin film solar cells, enabling the deposition of semiconductor layers at lower temperatures.

9.3.8   Ion Beam Assisted Deposition (IBAD)

IBAD combines ion bombardment with thin film deposition, enhancing adhesion, density, and optical properties of coatings.

Example: Optical Filters

IBAD is utilized in the fabrication of optical filters, where improved film properties contribute to enhanced performance.

Part V

Advanced Topics

Chapter 10

Quantum Optics

10.1 Quantum Mechanical Aspects

Quantum optics delves into the intersection of optics and quantum mechanics, exploring the behavior of light and its interaction with matter at the quantum level. Understanding quantum mechanical aspects is crucial for exploring phenomena beyond classical optics.

10.1.1 Wave-Particle Duality

The wave-particle duality of light is a fundamental concept in quantum optics, where light exhibits both wave and particle properties.

Example: Young's Double-Slit Experiment

Young's double-slit experiment demonstrates wave-like interference patterns when light passes through two slits, supporting the wave nature of light.

10.1.2 Quantum States of Light

Quantum optics introduces the concept of quantized states of light, represented by photons, each possessing discrete energy levels.

Example: Photon Counting

Photon counting experiments reveal the discrete nature of light, highlighting the quantized energy levels associated with individual photons.

10.1.3   Coherent and Incoherent Light

Coherent and incoherent light sources exhibit different characteristics in quantum optics, influencing interference and correlation phenomena.

Example: Laser Light Coherence

Laser light is an example of coherent light, where photons maintain a fixed phase relationship, leading to interference patterns.

10.1.4   Quantum Entanglement

Quantum entanglement is a phenomenon where the states of two or more particles become correlated, even when separated by large distances.

Example: EPR Paradox

The Einstein-Podolsky-Rosen (EPR) paradox illustrates quantum entanglement, challenging classical notions of locality and realism.

10.1.5   Quantum Superposition

Quantum superposition allows particles to exist in multiple states simultaneously, leading to unique optical effects.

Example: Schrödinger's Cat

Schrödinger's cat is a thought experiment highlighting quantum superposition, where a cat can be both alive and dead until observed.

10.1.6 Quantum Interference

Quantum interference occurs when multiple quantum paths contribute to the probability amplitude of a particle.

Example: Mach-Zehnder Interferometer

The Mach-Zehnder interferometer demonstrates quantum interference, where the superposition of paths influences the final outcome.

10.1.7 Quantum Optics in Information Processing

Quantum optics plays a crucial role in quantum information processing, including quantum computing and quantum communication.

Example: Quantum Key Distribution

Quantum key distribution utilizes quantum properties to secure communication channels, ensuring the privacy of transmitted information.

10.1.8 Applications in Quantum Technologies

Quantum optics applications extend to quantum technologies, impacting fields such as sensing, imaging, and cryptography.

Example: Quantum Sensing

Quantum sensors leverage quantum properties for enhanced sensitivity, enabling precise measurements in various scientific and industrial applications.

10.2 Photon Statistics

Photon statistics in quantum optics deal with the distribution of photons in various states, providing insights into the behavior of light at the quantum level. Understanding photon statistics is crucial for applications such as quantum communication and quantum information processing.

10.2.1   Introduction to Photon Statistics

Photon statistics characterize the probability distribution of photons in different quantum states, determining the randomness or coherence of light.

Example: Thermal Light Source

A thermal light source exhibits Poissonian statistics, where the number of photons in a given time interval follows a Poisson distribution.

10.2.2   Poissonian Photon Statistics

Poissonian photon statistics describe the distribution of photons in a classical, independent, and random manner.

Example: Weak Coherent Light

Weak coherent light, such as that from a laser with low intensity, often exhibits Poissonian statistics, where photon arrivals are independent events.

10.2.3   Super-Poissonian Photon Statistics

Super-Poissonian statistics involve photon distributions with higher variance than a Poisson distribution, indicating photon bunching.

Example: Quantum Dots

Quantum dots can emit photons in a correlated manner, leading to super-Poissonian statistics due to the phenomenon of photon bunching.

10.2.4   Sub-Poissonian Photon Statistics

Sub-Poissonian statistics occur when the photon distribution has a lower variance than a Poisson distribution, indicating anti-bunching.

Example: Single Photon Sources

Single photon sources, like certain quantum emitters, often exhibit sub-Poissonian statistics, with photons arriving in a more regular and anti-correlated manner.

10.2.5 Glauber's Photon Correlation Function

Glauber's photon correlation function quantifies the correlation between photon arrivals, providing a powerful tool for analyzing photon statistics.

Example: Hanbury Brown and Twiss Experiment

The Hanbury Brown and Twiss experiment utilizes Glauber's correlation function to measure the degree of photon bunching or anti-bunching in light emitted from distant stars.

10.2.6 Applications in Quantum Information

Photon statistics play a crucial role in quantum information processing, influencing the design and performance of quantum communication systems.

Example: Quantum Key Distribution

In quantum key distribution, understanding photon statistics is essential for ensuring secure communication channels through the detection of eavesdropping attempts.

10.2.7 Quantum Optics Experiments

Various experiments in quantum optics involve manipulating photon statistics to observe quantum phenomena, advancing our understanding of light at the quantum level.

Example: Hong-Ou-Mandel Experiment

The Hong-Ou-Mandel experiment explores photon bunching and anti-bunching effects, demonstrating the quantum nature of light through interference.

10.3   Quantum Entanglement

Quantum entanglement is a profound phenomenon in quantum optics where particles become interconnected, sharing correlated states regardless of the distance between them. Understanding quantum entanglement is fundamental for various quantum technologies and experiments.

10.3.1   Basics of Quantum Entanglement

Quantum entanglement involves the creation of entangled pairs of particles, such as photons or electrons, with properties linked in a way that the state of one particle instantly influences the state of the other.

Example: Entangled Photon Pairs

In quantum optics experiments, entangled photon pairs can be created through processes like spontaneous parametric down-conversion, leading to correlated polarization states.

10.3.2   Bell's Theorem and Entanglement

Bell's theorem establishes that certain predictions of quantum mechanics are incompatible with classical physics, and entangled particles violate Bell inequalities, showcasing the non-local nature of quantum entanglement.

Example: Bell Inequality Violation

Experimental tests of Bell inequalities confirm the violation, demonstrating the entangled nature of particles and challenging classical notions of local realism.

10.3.3 Quantum Entanglement and Information

Quantum entanglement is integral to quantum information processing, enabling the creation of entangled qubits for applications in quantum computing and quantum communication.

Example: Quantum Teleportation

Quantum teleportation relies on entangled particles to transmit quantum states between distant locations, showcasing the potential for secure quantum communication.

10.3.4 Entanglement in Quantum Experiments

Various quantum experiments exploit entanglement to observe unique phenomena, including quantum interference and quantum correlations.

Example: Aspect Experiment

The Aspect experiment demonstrates entanglement through measuring correlations between entangled particles' polarizations, providing experimental validation of quantum entanglement.

10.3.5 Quantum Entanglement in Photons

Entanglement in photons plays a crucial role in quantum optics, with applications ranging from quantum key distribution to quantum imaging.

Example: Quantum Key Distribution

Quantum key distribution protocols, such as the BBM92 protocol, leverage entangled photon pairs to establish secure communication channels resistant to eavesdropping.

10.3.6   Applications in Quantum Cryptography

Quantum entanglement serves as the foundation for quantum cryptographic protocols, ensuring the security of quantum communication channels.

Example: E91 Protocol

The E91 protocol utilizes entangled particles to create a secret key, forming the basis for secure quantum communication in quantum cryptography.

10.3.7   Future Directions in Entanglement Research

Ongoing research explores new frontiers in quantum entanglement, aiming to harness its unique properties for advanced quantum technologies.

Example: Quantum Repeaters

Quantum repeaters are a promising avenue for extending the range of entangled particle transmission, overcoming limitations in quantum communication distances.

Chapter 11

Nonlinear Optics

11.1 Introduction to Nonlinear Effects

Nonlinear optics explores the interaction of intense light with matter, leading to phenomena that cannot be explained by linear optics. Understanding nonlinear effects is crucial for various applications, from laser technology to optical signal processing.

11.1.1 Overview of Nonlinear Optics

Nonlinear optics studies the response of materials to high-intensity light, where the polarization of the medium becomes dependent on the intensity of the incident light.

Example: Second Harmonic Generation

In second harmonic generation, high-intensity light interacts with a nonlinear crystal, resulting in the generation of a new frequency at twice the input frequency.

11.1.2   Nonlinear Susceptibility

Nonlinear susceptibility $(\chi^{(n)})$ characterizes the nonlinear response of a material to an electric field and is crucial for understanding the strength of nonlinear effects.

Example: Third-Order Nonlinear Susceptibility

Third-order nonlinear susceptibility is often associated with phenomena like four-wave mixing, where four photons interact to generate new frequencies.

11.1.3   Kerr Effect

The Kerr effect is a nonlinear phenomenon where the refractive index of a material changes with the intensity of the incident light.

Example: Self-Focusing

Self-focusing, a consequence of the Kerr effect, occurs when the refractive index change causes light to focus on itself, leading to beam collapse.

11.1.4   Four-Wave Mixing

Four-wave mixing is a nonlinear process where four input waves interact to generate new frequency components.

Example: Optical Parametric Amplification

Optical parametric amplification involves four-wave mixing to amplify an input signal, enabling the creation of new frequencies.

11.1.5   Solitons

Solitons are stable, self-reinforcing wave packets that can propagate without changing shape, a phenomenon arising from a balance between nonlinear and dispersive effects.

Example: Optical Solitons

In optical communications, optical solitons can transmit information over long distances without distortion, overcoming dispersion effects.

11.1.6 Applications of Nonlinear Optics

Nonlinear optics has diverse applications, from laser frequency conversion to ultrafast optical signal processing.

Example: Frequency Doubling in Lasers

Frequency doubling, a nonlinear process, is used to convert the frequency of laser light for applications like green laser pointers.

11.1.7 Challenges in Nonlinear Optics

Despite its potential, nonlinear optics poses challenges, including material damage at high intensities and the need for precise control of nonlinear effects.

Example: Material Damage in Intense Fields

In high-intensity applications, material damage due to nonlinear effects can limit the practical implementation of nonlinear optics.

11.1.8 Emerging Trends in Nonlinear Optics

Ongoing research explores novel materials and techniques to enhance nonlinear effects and develop new applications.

Example: Nonlinear Photonic Crystal Fibers

Nonlinear photonic crystal fibers are being investigated for their potential to control and enhance nonlinear effects in compact and versatile platforms.

11.2 Second Harmonic Generation

Second harmonic generation (SHG) is a fascinating nonlinear optical process where a material generates light at double the frequency of the incident light. This phenomenon is essential in various applications, from laser technology to biological imaging.

11.2.1 Principle of Second Harmonic Generation

In SHG, a nonlinear medium interacts with a high-intensity laser beam, causing the generation of a new frequency at exactly twice the input frequency.

$$\omega_{\text{SH}} = 2\omega_{\text{fundamental}} \tag{11.1}$$

This equation represents the relationship between the second harmonic frequency (ω_{SH}) and the fundamental frequency ($\omega_{\text{fundamental}}$).

Example: SHG in a Nonlinear Crystal

Consider a crystal with a second-order nonlinear susceptibility ($\chi^{(2)}$). The induced polarization ($P^{(2)}$) due to the incident electric field (E) is given by:

$$P^{(2)} = \epsilon_0 \chi^{(2)} E^2 \tag{11.2}$$

Here, ϵ_0 is the vacuum permittivity.

11.2.2 Phase Matching in SHG

Phase matching is crucial for efficient SHG. It ensures that the generated second harmonic wave is in phase with the fundamental wave, maximizing the constructive interference.

Example: Birefringent Phase Matching

Birefringent phase matching involves selecting a crystal orientation where the refractive indices for the fundamental and second harmonic waves match, facilitating phase matching.

11.2.3 Applications of SHG

SHG finds applications in diverse fields, including laser sources, medical imaging, and material characterization.

Example: SHG in Biomedical Imaging

In biomedical imaging, SHG is utilized to visualize structures like collagen fibers, providing valuable information in fields such as microscopy.

11.2.4 Challenges in SHG

Despite its utility, SHG faces challenges such as low conversion efficiency and the need for precise phase matching.

Example: Temperature Sensitivity

The efficiency of SHG can be temperature-dependent, requiring control measures to maintain stable operation.

11.2.5 Emerging Trends in SHG

Ongoing research focuses on improving SHG efficiency, exploring new materials, and developing advanced techniques for enhanced applications.

Example: Nonlinear Metasurfaces

Nonlinear metasurfaces are investigated for their potential in controlling and enhancing SHG through engineered nanostructures.

11.3 Applications in Nonlinear Optics

Nonlinear optics plays a crucial role in various technological advancements, offering a wide range of applications that exploit the unique properties of materials under intense light. This section explores some notable applications and their underlying principles.

11.3.1   Harmonic Generation

One of the fundamental applications of nonlinear optics is harmonic generation, including second harmonic generation (SHG) and third harmonic generation (THG). These processes find applications in laser sources, microscopy, and material characterization.

Example: SHG in Laser Systems

In laser systems, SHG is employed to generate coherent light at double the frequency, providing a versatile tool for applications such as laser surgery and material processing.

Example: THG in Microscopy

THG is utilized in microscopy to achieve three-photon excitation, enabling deeper imaging in biological samples without damaging surrounding tissues.

11.3.2   Parametric Amplification

Parametric amplification involves the generation of a new signal through the nonlinear interaction of a pump wave with a material. This process is essential in creating tunable and efficient amplifiers.

Example: Optical Parametric Oscillator

The optical parametric oscillator (OPO) is a widely used device in parametric amplification. It can generate coherent light across a broad range of wavelengths, making it valuable in spectroscopy and imaging.

11.3.3   Soliton Propagation

Nonlinear optics supports the generation and propagation of solitons—localized wave packets that maintain their shape during transmission. Solitons find applications in optical communication systems.

Example: Soliton Communication

In optical fiber communication, solitons are employed to transmit information over long distances with minimal distortion, enhancing the efficiency of data transmission.

11.3.4 Nonlinear Optical Materials

Advancements in nonlinear optical materials have paved the way for innovative applications. Materials with high nonlinear susceptibilities are crucial for efficient nonlinear optical processes.

Example: Nonlinear Photonic Crystals

Nonlinear photonic crystals are designed to exhibit enhanced nonlinear effects, enabling applications in frequency conversion and signal processing.

11.3.5 Quantum Information Processing

Nonlinear optics contributes to the field of quantum information processing, where nonlinear interactions are harnessed for quantum computing and communication.

Example: Entangled Photon Generation

Nonlinear processes, such as spontaneous parametric down-conversion, are used to generate entangled photon pairs, a key resource in quantum communication protocols.

Part VI

Optics in Technology

Chapter 12

Optics in Imaging

12.1 Image Formation

Image formation is a fundamental aspect of optics, crucial in various technologies, especially imaging systems. This section explores the principles of image formation, the role of optical components, and relevant formulas.

12.1.1 Basic Concepts

Image formation involves the creation of a visual representation of an object using optical systems. Key concepts include object distance (d_o), image distance (d_i), and focal length (f).

The lens formula relates these parameters:

$$\frac{1}{f} = \frac{1}{d_o} + \frac{1}{d_i} \tag{12.1}$$

12.1.2 Lens Systems

The image formed by a lens depends on the type of lens and the position of the object. Convex lenses converge light, forming real or virtual images. Concave lenses diverge light, producing only virtual images.

Example: Convex Lens Image Formation

Consider a convex lens with an object placed beyond its focal point. The resulting image is real, inverted, and located on the opposite side of the lens.

12.1.3   Camera Systems

Modern cameras utilize lenses and image sensors to capture and record images. Understanding image formation is crucial for optimizing camera performance.

Example: Camera Aperture and Depth of Field

The size of the camera aperture affects the depth of field. A wider aperture (lower f-number) results in a shallower depth of field, impacting the focus range in an image.

12.1.4   Microscope Systems

Microscopes magnify small objects by forming highly magnified images. Optical elements in microscopes contribute to the clarity and resolution of the final image.

Example: Objective Lens Magnification

The magnification (M) of a microscope objective lens is given by the formula:

$$M = \frac{d_i}{d_o} \tag{12.2}$$

12.1.5   Telescope Systems

Telescopes are designed to observe distant objects. The combination of lenses and mirrors in telescopes determines their magnification and field of view.

Example: Telescope Eyepiece

The eyepiece of a telescope further magnifies the image formed by the objective lens or mirror, allowing detailed observations of celestial objects.

12.1.6 Holography

Holography is an advanced imaging technique that captures both the intensity and phase of light. It enables the creation of three-dimensional holograms.

Example: Holographic Interference Pattern

In holography, the interference pattern between reference and object waves is recorded. Reconstructing this pattern produces a realistic three-dimensional image.

12.2 Imaging Systems

Imaging systems play a pivotal role in capturing and reproducing visual information. This section delves into the principles behind imaging systems, covering topics such as resolution, aberrations, and the impact of different optical components.

12.2.1 Resolution in Imaging

The resolution of an imaging system determines its ability to distinguish details. Two essential components are considered: spatial resolution and contrast resolution.

Spatial Resolution

Spatial resolution is the ability to distinguish between closely spaced objects. It is influenced by factors such as the size of the detector pixels and the optical system's resolving power.

$$\text{Spatial Resolution} = \frac{\text{Field of View}}{\text{Number of Pixels}} \tag{12.3}$$

Contrast Resolution

Contrast resolution refers to the system's ability to distinguish between objects with similar intensities. It is affected by factors like the dynamic range of the detector and the modulation transfer function.

$$\text{Contrast Resolution} = \frac{\text{Contrast of Objects}}{\text{Contrast Sensitivity}} \tag{12.4}$$

12.2.2   Aberrations in Imaging

Aberrations are deviations from ideal imaging behavior and can affect image quality. Common aberrations include chromatic aberration, spherical aberration, and coma.

Chromatic Aberration

Chromatic aberration results from the dispersion of different wavelengths of light. It can be minimized using achromatic lenses that combine multiple lens elements.

$$f(\lambda) = \frac{f}{n(\lambda)} \tag{12.5}$$

Spherical Aberration

Spherical aberration occurs when light rays passing through the periphery of a lens focus at different points than those passing through the center. It can be reduced using aspheric lens designs.

$$f(r) = \frac{f}{1 + \frac{1}{2}\left(\frac{r}{R}\right)^2} \tag{12.6}$$

12.2.3   Optical Components in Imaging

Various optical components contribute to the functionality of imaging systems. Lenses, mirrors, and filters play crucial roles in shaping the image.

Example: Lens System for Microscopy

In microscopy, a combination of objective and eyepiece lenses contributes to the magnification and clarity of the image. The total magnification is the product of the magnifications of the individual lenses.

$$\text{Total Magnification} = \text{Magnification}_{\text{Objective}} \times \text{Magnification}_{\text{Eyepiece}} \quad (12.7)$$

12.3 Image Processing in Optics

Image processing plays a crucial role in enhancing and analyzing images captured through optical systems. This section explores various image processing techniques, their mathematical foundations, and practical applications.

12.3.1 Introduction to Image Processing

Image processing involves manipulating an image to extract valuable information or improve its visual quality. It encompasses operations such as filtering, enhancement, and segmentation.

Image Representation

Images can be represented as matrices, where each element corresponds to a pixel value. For a grayscale image, the matrix is two-dimensional, while for a colored image, it is three-dimensional.

$$\text{Image} = \begin{bmatrix} I_{1,1} & I_{1,2} & \cdots & I_{1,N} \\ I_{2,1} & I_{2,2} & \cdots & I_{2,N} \\ \vdots & \vdots & \ddots & \vdots \\ I_{M,1} & I_{M,2} & \cdots & I_{M,N} \end{bmatrix} \quad (12.8)$$

12.3.2   Image Filtering

Filtering is a fundamental operation in image processing that involves modifying pixel values based on a specified kernel or filter. Common filters include blurring, sharpening, and edge detection filters.

Blurring Filter

A blurring filter is often used to reduce noise or smooth an image. The formula for a simple averaging filter is given by:

$$\text{Output}_{i,j} = \frac{1}{9}\left(\sum_{m=1}^{3}\sum_{n=1}^{3}\text{Input}_{i+m-1,j+n-1}\right) \tag{12.9}$$

Edge Detection Filter

An edge detection filter highlights abrupt changes in pixel values, emphasizing edges in an image. One such filter is the Sobel operator:

$$\text{Gradient}_x = \begin{bmatrix} -1 & 0 & 1 \\ -2 & 0 & 2 \\ -1 & 0 & 1 \end{bmatrix}, \quad \text{Gradient}_y = \begin{bmatrix} -1 & -2 & -1 \\ 0 & 0 & 0 \\ 1 & 2 & 1 \end{bmatrix} \tag{12.10}$$

12.3.3   Image Enhancement

Image enhancement techniques aim to improve the visual quality of an image by adjusting its contrast, brightness, or color balance.

Contrast Stretching

Contrast stretching expands the range of pixel values in an image to utilize the full dynamic range. The formula for contrast stretching is:

$$\text{Output}_{i,j} = \frac{\text{Input}_{i,j} - \text{Min}_{\text{Input}}}{\text{Max}_{\text{Input}} - \text{Min}_{\text{Input}}} \times (\text{Max}_{\text{Output}} - \text{Min}_{\text{Output}}) + \text{Min}_{\text{Output}}$$

$$\tag{12.11}$$

Chapter 13

Optics in Communication

13.1 Fiber Optics

Fiber optics revolutionized communication systems, enabling the transmission of information over long distances with minimal signal loss. This section delves into the principles of fiber optics, exploring the key concepts, components, and applications.

13.1.1 Introduction to Fiber Optics

Fiber optics involves the transmission of light through thin, flexible strands of glass or plastic fibers. The fundamental principle is based on total internal reflection, where light is trapped within the core of the fiber, allowing it to travel over extended distances.

Basic Components of Fiber Optic System

A typical fiber optic system consists of three main components:

1. **Optical Transmitter:** Converts electrical signals into optical signals for transmission.

2. **Optical Fiber:** Carries the optical signals over long distances.

3. **Optical Receiver:** Converts received optical signals back into electrical signals.

13.1.2   Propagation of Light in Optical Fibers

The propagation of light in optical fibers is governed by the principles of geometric optics. The numerical aperture (NA) of a fiber and the angle of incidence determine the acceptance angle and the maximum angle for total internal reflection.

$$\text{NA} = \sqrt{n_1^2 - n_2^2} \tag{13.1}$$

where n_1 is the refractive index of the core and n_2 is the refractive index of the cladding.

13.1.3   Types of Optical Fibers

There are several types of optical fibers designed for various applications. Common types include:

- Single-Mode Fiber (SMF)

- Multi-Mode Fiber (MMF)

- Graded-Index Fiber

Each type has its own characteristics and is suitable for different communication scenarios.

13.1.4   Working Example: Signal Loss Calculation

Consider a single-mode fiber with a length of 10 km. Calculate the signal loss given an attenuation coefficient of $0.2\,\text{dB/km}$.

$$\text{Total Loss} = \text{Attenuation Coefficient} \times \text{Length} \tag{13.2}$$

Substitute the given values to find the total signal loss.

13.1.5 Numerical Example: Bandwidth-Distance Product

Determine the bandwidth-distance product for a multi-mode fiber with a bandwidth of 500 MHz·km and a distance of 2 km.

$$\text{Bandwidth-Distance Product} = \text{Bandwidth} \times \text{Distance} \qquad (13.3)$$

Substitute the given values to find the bandwidth-distance product.

13.2 Optical Communication Systems

Optical communication systems form the backbone of modern telecommunications, facilitating high-speed data transfer over long distances using light signals. This section explores the fundamental principles, components, and applications of optical communication.

13.2.1 Introduction to Optical Communication

Optical communication relies on the transmission of information through optical fibers using light signals. It has revolutionized the way data is transmitted, offering higher bandwidth and lower signal loss compared to traditional copper-based systems.

Basic Components of Optical Communication System

An optical communication system comprises several key components:

1. **Optical Transmitter:** Converts electrical signals into optical signals.

2. **Optical Fiber:** Transmits optical signals over long distances.

3. **Optical Receiver:** Converts received optical signals back into electrical signals.

13.2.2 Working Principles of Optical Communication

The working principles involve the modulation of light signals to encode information. Common modulation techniques include amplitude modulation (AM), frequency modulation (FM), and phase modulation (PM).

$$P(t) = P_0 + A \cos(2\pi f t + \phi) \qquad (13.4)$$

where $P(t)$ is the modulated optical power, P_0 is the average optical power, A is the amplitude, f is the frequency, t is time, and ϕ is the phase.

13.2.3 Types of Optical Communication Systems

There are various types of optical communication systems catering to different needs:

- **Point-to-Point Communication:** Direct communication between two points.

- **Fiber Optic Networks:** Complex networks transmitting data over multiple nodes.

- **Free-Space Optical Communication:** Communication through free space using lasers.

13.2.4 Working Example: Modulation of Optical Signal

Consider an optical communication system using amplitude modulation with an average optical power of $P_0 = 10\,\text{mW}$, amplitude $A = 5\,\text{mW}$, frequency $f = 1\,\text{GHz}$, and phase $\phi = \pi/4$. Express the modulated optical power $P(t)$.

$$P(t) = 10 + 5\cos\left(2\pi \times 10^9 \times t + \frac{\pi}{4}\right) \qquad (13.5)$$

Substitute the given values to find the expression for $P(t)$.

13.2.5 Numerical Example: Optical Fiber Link Budget

Calculate the link budget for an optical fiber communication system with a transmitter power of $10\,\text{dBm}$, a receiver sensitivity of $-30\,\text{dBm}$, and a fiber attenuation of $0.2\,\text{dB/km}$ over a distance of $50\,\text{km}$.

$$\text{Link Budget} = \text{Transmitter Power} - \text{Fiber Attenuation} \times \text{Distance} - \text{Receiver Sensitivity} \tag{13.6}$$

$$= 10 - (0.2 \times 50) - (-30) \tag{13.7}$$

Calculate the link budget for the given values.

13.3 Emerging Technologies

The field of optics is dynamic, with continuous advancements leading to the emergence of new technologies. This section explores some of the cutting-edge technologies shaping the future of optics.

13.3.1 Metamaterials and Transformation Optics

Metamaterials are artificial materials engineered to exhibit properties not found in nature. In optics, metamaterials are designed to manipulate electromagnetic waves, enabling unconventional control over light. Transformation optics, a branch of metamaterial optics, focuses on bending and redirecting light to achieve novel optical effects.

Application Example: Invisibility Cloaks

Metamaterials can be used to create invisibility cloaks that render objects invisible by bending incident light around them. The design involves precise control of refractive indices to achieve the desired cloaking effect.

13.3.2 Plasmonics and Nanophotonics

Plasmonics deals with the interaction between electromagnetic field and free electrons in a metal. This field has applications in nanophotonics, where light is manipulated on the nanoscale. Plasmonic structures enable the confinement of light to dimensions much smaller than the wavelength.

Working Example: Surface Plasmon Resonance

Consider a plasmonic sensor utilizing surface plasmon resonance. The resonance condition is given by the formula:

$$\sin(\theta) = \frac{n_1}{n_2} \tag{13.8}$$

where θ is the angle of incidence, and n_1 and n_2 are the refractive indices of the surrounding medium and the metal, respectively.

13.3.3 Quantum Information and Quantum Optics

The integration of quantum mechanics with optics has given rise to quantum optics, exploring the quantum nature of light and its interaction with matter. Quantum information processing leverages quantum states for information storage and computation.

Numerical Example: Quantum Entanglement

Consider two entangled photons in a quantum optics experiment. The state of the entangled pair is given by the Bell state:

$$|\Psi\rangle = \frac{1}{\sqrt{2}}(|H\rangle \otimes |V\rangle - |V\rangle \otimes |H\rangle) \tag{13.9}$$

where $|H\rangle$ and $|V\rangle$ represent horizontal and vertical polarization states.

Part VII

Appendices

Appendix A

Glossary of Optical Terms

This glossary provides concise definitions of key optical terms to aid readers in understanding the terminology used in the field of optics.

Achromatic Lens

An achromatic lens is designed to minimize chromatic aberration, resulting in minimal color dispersion. It typically consists of multiple lens elements made from different types of glass.

Brewster's Angle

Brewster's angle is the angle of incidence at which light with a particular polarization is perfectly transmitted through a transparent dielectric surface, with no reflection. It is given by the formula:

$$\tan(\theta_B) = \frac{n_2}{n_1} \tag{A.1}$$

where θ_B is Brewster's angle, and n_1 and n_2 are the refractive indices of the incident and transmitted media, respectively.

Critical Angle

The critical angle is the angle of incidence beyond which total internal reflection occurs. It is given by Snell's Law:

$$\sin(\theta_c) = \frac{n_2}{n_1} \tag{A.2}$$

where θ_c is the critical angle, and n_1 and n_2 are the refractive indices of the first and second media.

Diffractive Optics

Diffractive optics involves the use of diffraction gratings or elements to manipulate light. Diffractive optical elements (DOEs) can be used for beam shaping, wavefront correction, and other applications.

Eikonal Equation

The eikonal equation describes the evolution of the phase of a wavefront. For a monochromatic wave, it is given by:

$$|\nabla \phi|^2 = n^2 - 1 \tag{A.3}$$

where ϕ is the phase of the wave and n is the refractive index.

Fresnel Equations

The Fresnel equations describe the reflection and transmission of light at an interface between two media. For normal incidence, they are given by:

$$R = \left(\frac{n_1 - n_2}{n_1 + n_2} \right)^2 \tag{A.4}$$

$$T = 1 - R \tag{A.5}$$

where R is the reflectance and T is the transmittance.

Gaussian Beam

A Gaussian beam is a beam of monochromatic light whose electric field amplitude follows a Gaussian distribution across its transverse section.

Huygens-Fresnel Principle

The Huygens-Fresnel principle states that each point on a wavefront acts as a secondary source of spherical wavelets, and the sum of these wavelets determines the shape of the wavefront at a later time.

Image Plane

The image plane is the plane in which a focused image is formed by an optical system. It is the plane where the image is most sharply defined.

Joule Heating

Joule heating refers to the heat generated in a medium due to the absorption of electromagnetic radiation, such as light. It is characterized by the rate of heat production per unit volume.

Kerr Effect

The Kerr effect is a change in the refractive index of a material induced by an electric field. It is a nonlinear optical phenomenon.

Laser Cavity

The laser cavity is the optical resonator within a laser where light is amplified through the process of stimulated emission. It typically consists of mirrors at the ends.

Microscope Objective

The microscope objective is the lens closest to the object being observed in a microscope. It provides the primary magnification of the specimen.

Numerical Aperture

Numerical aperture is a dimensionless number that characterizes the range of angles over which the system can accept or emit light. It is defined as:

$$\text{NA} = n \cdot \sin(\theta) \tag{A.6}$$

where n is the refractive index of the medium and θ is the half-angle of the maximum cone of light that can enter or exit the optical system.

Optical Density

Optical density, or absorbance, is a measure of how strongly a substance absorbs light at a particular wavelength. It is defined as the logarithm of the reciprocal of transmittance:

$$\text{OD} = -\log_{10}(T) \tag{A.7}$$

where OD is the optical density and T is the transmittance.

Polarization State

The polarization state of light describes the orientation of its electric field vector. Common polarization states include linear, circular, and elliptical polarization.

Quantum Efficiency

Quantum efficiency is a measure of the effectiveness of a detector in converting incident photons into electrical signals. It is defined as the ratio of the number of photoelectrons emitted to the number of incident photons.

Rayleigh Scattering

Rayleigh scattering is the elastic scattering of light or other electromagnetic radiation by particles much smaller than the wavelength of the radiation. It is responsible for the blue color of the sky.

Snell's Law

Snell's Law describes the relationship between the angles of incidence and refraction when light passes through an interface between two media. It is given by:

$$n_1 \cdot \sin(\theta_1) = n_2 \cdot \sin(\theta_2) \tag{A.8}$$

where n_1 and n_2 are the refractive indices of the two media, and θ_1 and θ_2 are the angles of incidence and refraction, respectively.

Total Internal Reflection

Total internal reflection occurs when light traveling in a medium strikes the interface with another medium at an angle greater than the critical angle. It results in the complete reflection of the light within the first medium.

Ultraviolet (UV) Light

Ultraviolet light is electromagnetic radiation with a wavelength shorter than that of visible light but longer than X-rays. It is often divided into UVA, UVB, and UVC regions based on wavelength.

Vectorial Optics

Vectorial optics deals with the description of light as an electromagnetic wave, considering both electric and magnetic field vectors. It provides a more comprehensive understanding of optical phenomena.

Wavefront Aberration

Wavefront aberration refers to deviations of a wavefront from its ideal shape. It is characterized by optical imperfections such as defocus, astigmatism, and coma.

Xenon Lamp

A xenon lamp is a gas-discharge lamp that produces a bright and broad spectrum of light, making it suitable for various applications, including optical systems and projectors.

Young's Double-Slit Experiment

Young's double-slit experiment demonstrates the interference of light waves. It involves shining light through two closely spaced slits, leading to an interference pattern on a screen.

Zeeman Effect

The Zeeman effect is the splitting of spectral lines in the presence of a magnetic field. It is a result of the interaction between magnetic moments associated with the electron's intrinsic spin and its orbital motion.

This glossary provides a comprehensive overview of key optical terms, offering readers a valuable reference for understanding the language of optics. As the field continues to advance, this glossary serves as a foundation for exploring new concepts and technologies.

Appendix B

Mathematical Formulas

This appendix provides a collection of essential mathematical formulas relevant to optics. These formulas cover a range of topics, from basic trigonometry to advanced mathematical expressions commonly encountered in optics.

Trigonometric Identities

The following trigonometric identities are frequently used in optics:

$$\sin^2(\theta) + \cos^2(\theta) = 1 \tag{B.1}$$

$$\tan(\theta) = \frac{\sin(\theta)}{\cos(\theta)} \tag{B.2}$$

$$\cos(2\theta) = \cos^2(\theta) - \sin^2(\theta) \tag{B.3}$$

These identities form the basis for understanding wave propagation, interference, and diffraction phenomena.

Matrix Representation of Optical Systems

Matrix methods are commonly employed to describe the action of optical elements on light rays. The matrix representation of a lens with focal length f and refractive index n is given by:

$$\begin{bmatrix} 1 & 0 \\ -\frac{1}{f} & \frac{1}{n} \end{bmatrix} \tag{B.4}$$

This matrix can be used to calculate the overall effect of a sequence of optical elements.

Fourier Transform in Optics

The Fourier transform is crucial in analyzing spatial frequency components of optical signals. For a function $f(x)$, its Fourier transform $F(u)$ is defined by:

$$F(u) = \int_{-\infty}^{\infty} f(x) \cdot e^{-i2\pi u x} \, dx \tag{B.5}$$

This allows the decomposition of optical signals into their frequency components.

Maxwell's Equations

Maxwell's equations describe the behavior of electromagnetic fields. In the absence of free charges and currents, they are given by:

$$\nabla \cdot \mathbf{E} = 0 \tag{B.6}$$

$$\nabla \cdot \mathbf{B} = 0 \tag{B.7}$$

$$\nabla \times \mathbf{E} = -\frac{\partial \mathbf{B}}{\partial t} \tag{B.8}$$

$$\nabla \times \mathbf{B} = \mu_0 \mathbf{J} + \mu_0 \epsilon_0 \frac{\partial \mathbf{E}}{\partial t} \tag{B.9}$$

These equations form the foundation of classical optics and wave optics.

Fresnel Diffraction Integral

The Fresnel diffraction integral describes the diffraction pattern formed when light passes through a small aperture. It is given by:

$$U(P) = \frac{1}{i\lambda} \iint_\Sigma U(Q) \frac{e^{ikr}}{r} \cos(\theta) \, d\Sigma \qquad \text{(B.10)}$$

where $U(P)$ is the complex amplitude at point P, $U(Q)$ is the complex amplitude at the aperture point Q, k is the wave number, and r is the distance between Q and P.

Rayleigh-Sommerfeld Diffraction Integral

The Rayleigh-Sommerfeld diffraction integral is an extension of the Fresnel diffraction integral that includes near-field effects. It is expressed as:

$$U(P) = \frac{1}{i\lambda} \iint_\Sigma U(Q) \frac{e^{ikr}}{r} \cos(\theta) \, d\Sigma - \frac{ik}{2} \iint_\Sigma U(Q) \frac{e^{ikr}}{r^2} \, d\Sigma \qquad \text{(B.11)}$$

This integral provides a more accurate description of diffraction phenomena in the near field.

Quantum Mechanics in Optics

Quantum mechanics plays a crucial role in understanding the behavior of photons and matter at the quantum level. The wave function ψ satisfies the Schrödinger equation:

$$i\hbar \frac{\partial \psi}{\partial t} = -\frac{\hbar^2}{2m} \nabla^2 \psi + V\psi \qquad \text{(B.12)}$$

where $\hbar$ is the reduced Planck constant, m is the particle mass, V is the potential energy, and ∇^2 is the Laplacian operator.

Optical Fiber Communication

In optical fiber communication, the performance of a system is often characterized by the bit error rate (BER), defined as the ratio of the number of received bits in error to the total number of transmitted bits.

$$\text{BER} = \frac{\text{Number of Bits in Error}}{\text{Total Number of Transmitted Bits}} \qquad \text{(B.13)}$$

The BER is a critical parameter in assessing the quality of optical communication systems.

Geometrical Optics

Geometrical optics provides a simplified model for understanding the propagation of light in optical systems. Snell's Law is a fundamental formula in geometrical optics:

$$n_1 \sin(\theta_1) = n_2 \sin(\theta_2) \tag{B.14}$$

where n_1 and n_2 are the refractive indices of the media, and θ_1 and θ_2 are the angles of incidence and refraction, respectively.

These mathematical formulas cover a broad spectrum of topics in optics, ranging from classical wave optics to quantum optics and communication systems. The understanding of these formulas is crucial for researchers, engineers, and students working in the field of optics.

Appendix C

Optical Constants of Materials

This appendix presents a comprehensive compilation of optical constants for various materials commonly encountered in optics. Optical constants, such as refractive index (n) and extinction coefficient (k), play a crucial role in characterizing the interaction of light with materials.

Refractive Index (n)

The refractive index of a material is a fundamental optical property that influences the speed of light within the material. It is defined as the ratio of the speed of light in a vacuum to the speed of light in the material.

$$n = \frac{c}{v} \tag{C.1}$$

where: n is the refractive index, c is the speed of light in a vacuum (3.00×10^8 m/s), v is the speed of light in the material.

Extinction Coefficient (k)

The extinction coefficient quantifies how strongly a material absorbs light. It is related to the imaginary part of the complex refractive index.

$$k = \frac{4\pi}{\lambda} \cdot \text{Im}\{\sqrt{\varepsilon}\} \tag{C.2}$$

where: k is the extinction coefficient, λ is the wavelength of light, ε is the complex dielectric constant.

Sample Optical Constants

Here are the optical constants for selected materials at a specific wavelength (λ):

Material	$\lambda(\mu m)$	n	k	
				(C.3)
Silicon	0.633	3.886	0.019	(C.4)
Glass	0.55	1.5	0.0	(C.5)
Gold	0.55	$0.200 + 3.724i$	$2.464 + 1.771i$	(C.6)
Water	0.589	1.333	0.0	(C.7)
				(C.8)

These values are indicative and can vary based on the specific composition and conditions of the material. Researchers and engineers often consult databases or conduct experiments to obtain accurate optical constants for their applications.

Applications

The knowledge of optical constants is crucial in designing optical components, coatings, and devices. For example, the selection of materials with specific refractive indices is essential in lens design, while the absorption characteristics (k) are critical in designing filters and coatings.

Understanding how different materials interact with light allows for the optimization of optical systems in terms of performance and efficiency.